Applied Psychology
for Nurses

P. G. Wodehouse

Applied Psychology for Nurses

Copyright © 2020 Bibliotech Press
All rights reserved

The present edition is a reproduction of previous publication of this classic work. Minor typographical errors may have been corrected without note, however, for an authentic reading experience the spelling, punctuation, and capitalization have been retained from the original text.

ISBN: 978-1-64799-147-0

FOREWORD

This little book is the outgrowth of a conviction, strengthened by some years of experience with hundreds of supposedly normal young people in schools and colleges, confirmed by my years of training in a neurological hospital and months of work in a big city general hospital, that it is of little value to help some people back to physical health if they are to carry with them through a prolonged life the miseries of a sick attitude. As nurses I believe it is our privilege and our duty to work for health of body and health of mind as inseparable. Experience has proved that too often the physically ill patient (hitherto nervously well) returns from hospital care addicted to the illness-accepting attitude for which the nurse must be held responsible.

I conceive of it as possible that every well trained nurse in our country shall consider it an essential to her professional success to leave her patient imbued with the will to health and better equipped to attain it because the sick attitude has been averted, or if already present, has been treated as really and intelligently as the sick body. To this end I have dealt with the simple principles of psychology only as the nurse can immediately apply them.

The writer wishes to acknowledge her indebtedness for criticism of this work and for several definitions better than her own, in the chapters The Normal Mind and Variations From Normal Mental Processes, to Dr. Robert S. Carroll, who through the years of hospital training helped her to translate her collegiate psychology from fascinating abstract principles into the sustaining bread of daily life.

Mary F. Porter.
Asheville, N. C.,
August, 1921.

CONTENTS

CHAPTER I: What is Psychology? 1
CHAPTER II: Consciousness 8
CHAPTER III: Organs of Consciousness 18
CHAPTER IV: Relation of Mind and Body 22
CHAPTER V: The Normal Mind 27
CHAPTER VI : The Normal Mind (Continued) 37
CHAPTER VII: Psychology and Health 52
CHAPTER VIII: Variations from Normal Mental Processes 63
CHAPTER IX: Variations from Normal Mental Processes (Continued) 70
CHAPTER X: Attention the Root of Disease or Health Attitude 80
CHAPTER XI: Getting the Patient's Point of View 88
CHAPTER XII: The Psychology of the Nurse 99
CHAPTER XIII: The Psychology of the Nurse (Continued) ... 107
CHAPTER XIV: The Nurse of the Future 119

CHAPTER I
WHAT IS PSYCHOLOGY?

Wise men study the sciences which deal with the origins and development of animal life, with the structure of the cells, with the effect of various diseases upon the tissues and fluids of the body; they study the causes of the reactions of the body cells to disease germs, and search for the origin and means of extermination of these enemies to health. They study the laws of physical well-being. They seek for the chemical principles governing the reactions of digestive fluids to the foods they must transform into heat and energy. So the doctor learns to combat disease with science, and at the same time to apply scientific laws of health that he may fortify the human body against the invasion of harmful germs. Thus, eventually, he makes medicine itself less necessary.

But another science must walk hand in hand today with that of medicine; for doctors and nurses are realizing as never before the power of mind over body, and the hopelessness of trying to cure the one without considering the other. Hence psychology has come into her own as a recognized science of the mind, just as biology, histology, chemistry, pathology, and medicine are recognized sciences governing the body. As these are concerned with the "how" and "why" of life, and of the body reactions, so psychology is concerned with the "how" and "why" of conduct and of thinking. For as truly as every infectious disease is caused by a definite germ, just as truly has every action of man its adequate explanation, and every thought its definite origin. As we would know the laws of the sciences governing man's physical well-being that we might have body health, so we would know the laws of the mind and of its response to its world in order to attain and hold fast to mind health. Experience with patients soon proves to us nurses that the weal and woe of the one vitally affects the other.

"Psychology is the science of mental life, both of its phenomena and their conditions."

So William James took up the burden of proof some thirty years ago, and assured a doubting world of men and women that there were laws in the realm of mind as certain and dependable as those applying to the world of matter—men and women who were not at all sure they had any right to get near enough the center of things to see the wheels go round. But today thousands of people are trying to find out something of the way the mind is conceived, and to understand its workings. And many of us have in our impatient, hasty investigation, self-analytically taken our mental machines all to pieces and are trying effortfully to put them together again. Some of us have made a pretty bad mess of it, for we tore out the screws and pulled apart the adjustments so hastily and carelessly that we cannot now find how they fit. And millions of other machines are working wrong because the engineers do not know how to keep them in order, put them in repair, or even what levers operate them. So books must be written—books of directions.

If you can glibly recite the definition above, know and explain the meaning of "mental life," describe "its phenomena and their conditions," illustrating from real life; if you can do this, and prove that psychology is a science, i. e., an organized system of knowledge on the workings of the mind—not mere speculation or plausible theory—then you are a psychologist, and can make your own definitions. Indeed, the test of the value of a course such as this should be your ability, at its end, to tell clearly, in a few words of your own, what psychology is.

The word science comes from a Latin root, scir, the infinitive form, scire, meaning to know. So a science is simply the accumulated, tested knowledge, the proved group of facts about a subject, all that is known of that subject to date. Hence, if psychology is a science, it is no longer a thing of guesses or theories, but is a group ing of confirmed facts about the mind, facts proved in the psychology laboratory even as chemical facts are demonstrated in the chemical laboratory. Wherein psychology departs from facts which can be proved by actual experience or by accurate tests, it becomes metaphysics, and is beyond the realm of science; for metaphysics deals with the realities of the supermind, or the soul, and its relations to life, and death, and God. Physics, chemistry, biology have all in their day been merely speculative. They were bodies of theory which might prove true or might not. When they worked, by actually being tried out, they became bodies of accepted facts, and are today called sciences. In the same way the laws of the working of the mind have been tested, and a body of assured facts about it has taken its place with other sciences.

It must be admitted that no psychologist is willing to stop with the known and proved, but, when he has presented that, dips into the fascinations of the yet unknown, and works with promising theory, which tomorrow may prove to be science also. But we will first find what they have verified, and make that the safe foundation for our own understanding of ourselves and others.

What do we mean by "mental life"?—or, we might say, the science of the life of the mind. And what is mind?

But let us start our quest by asking first what reasons we have for being sure mind exists. We find the proof of it in consciousness, although we shall learn later that the activities of the mind may at times be unconscious. So where consciousness is, we know there is mind; but where consciousness is not, we must find whether it has been, and is only temporarily withdrawn, before we say "Mind is not here." And consciousness we might call awareness, or our personal recognition of being—awareness of me, and thee, and it. So we recognize mind by its evidences of awareness, i. e., by the body's reaction to stimuli; and we find mind at the very dawn of animal life.

Consciousness is evidenced in the protozoön, the simplest form in which animal life is known to exist, by what we call its response to stimuli. The protozoön has a limited power of self-movement, and will accept or reject certain environments. But while we see that mind expresses itself in consciousness as vague, as dubious as that of the protozoön, we find it also as clear, as definite, as far reaching as that of the statesman, the chemist, the philosopher. Hence, the "phenomena of mental life" embrace the entire realms of feeling, knowing, willing—not of man alone, but of all creatures.

In our study, however, we shall limit ourselves to the psychology of the human mind, since that concerns us vitally as nurses. Animal psychology, race psychology, comparative psychology are not within the realm of our practical needs in hospital life. We would know the workings of man's mind in disease and health. What are the instinctive responses to fear, as shown by babies and children and primitive races? What are the normal expressions of joy, of anger, or desire? What external conditions call forth these evidences? What are the acquired responses to the things which originally caused fear, or joy, or anger? How do grown-ups differ in their reactions to the same stimuli? Why do they differ? Why does one man walk firmly, with stern, set face, to meet danger? Why does another quake and run? Why does a third man approach it with a swagger, face it with a confident, reckless smile of defiance?

All these are legitimate questions for the psychologist. He will approach the study of man's mind by finding how his body acts— that is, by watching the phenomena of mental life—under various conditions; then he will seek for the "why" of the action. For we can only conclude what is in the mind of another by interpreting his expression of his thinking and feeling. We cannot see within his mind. But experience with ourselves and others has taught us that certain attitudes of body, certain shades of countenance, certain gestures, tones of voice, spontaneous or willed actions, represent anger or joy, impatience or irritability, stern control or poise of mind. We realize that the average man has learned to conceal his mental reactions from the casual observer at will. But if we see him at an unguarded moment, we can very often get a fair idea of his mental attitude. Through these outward expressions we are able to judge to some extent of the phenomena of his mental life. But let us list them from our own minds as they occur to us this work-a-day moment, then, later on, find what elements go to make up the present consciousness.

As I turn my thoughts inward at this instant I am aware of these mental impressions passing in review:

You nurses for whom I am writing.

The hospitals you represent.

What you already know or do not know along these lines.

A child calling on the street some distance away.

A brilliant sunshine bringing out the sheen of the green grass.

The unmelodious call of a flicker in the pine-tree, and a towhee singing in the distance.

A whistling wind bending the pines.

A desire to throw work aside and go for a long tramp.

A patient moving about overhead (she is supposed to be out for her walk, and I'm wondering why she is not).

The face and voice of an old friend whom I was just now called from my work to see.

The plan and details of my writing.

The face and gestures of my old psychology professor and the assembled class engaged in a tangling metaphysic discussion.

A cramped position.

Some loose hair about my face distracting me.

An engagement at 7.30.

A sharp resolve to stop wool-gathering and finish this chapter.

And yet, until I stopped to examine my consciousness, I was keenly aware only of the thoughts on psychology I was trying to put on paper.

But how shall we classify these various contents?

Some are emotion, i. e., feelings; others are intellect, i. e., thoughts; still others represent determination, i. e., volition or will.

There is nothing in this varied consciousness that will not be included in one or another of these headings. Let us group the contents for ourselves.

The nurses for whom I am writing:

A result of memory and of imagination (both intellect). A sense of kinship and interest in them (emotion). A determination that they must have my best (will, volition).

And so of the hospitals:

My memory of hospitals I have known, and my mental picture of yours made up from piecing together the memories of various ones, the recollection of the feelings I had in them, etc. (intellect).

6

What you already know.

Speculation (intellect), the speculation based on my knowledge of other schools (memory which is intellect). A desire (emotion) that all nurses should know psychology.

Child calling on street.

Recognition of sound (intellect) and pleasant perception of his voice (emotion).

Desire to throw work aside and go for a tramp on this gorgeous day.

Emotion, restrained by stronger emotion of interest in work at hand, and intellect, which tells me that this is a work hour—and will, which orders me to pay attention to duties at hand.

So all the phenomena of mental life are included in feelings, thoughts, and volitions which accompany every minute of my waking life, and probably invade secretly every second of my sleeping life.

The conditions of mental life—what are they?

In man and the higher animals the central nervous system, which, anatomy teaches us, consists of the brain and spinal cord. (In the lowest forms of animal life, a diffused nervous system located throughout the protoplasm.)
An external world.
A peripheral nervous system connecting the central nervous system with the outside world.
The sympathetic nervous system, provided to assure automatic workings of the vital functions of the body. These organs of the mind will be discussed in a later chapter.

CHAPTER II
CONSCIOUSNESS

We took a glimpse at random into the mental life of an adult consciousness, and found it very complicated, constantly changing. We found it packed with shifting material, which, on the surface, seemed to bear very little relation. We found reason, feeling, and will all interacting. We found nothing to indicate that a consciousness as simple as mere awareness might exist. We believe there might be such in the newborn babe, perhaps even in the baby a month old; but can we prove it? Let us look within again and see if there are not times of mere, bare consciousness in our own experience that give us the proof we need.

I have slept deeply all night. It is my usual waking time. Something from within or from without forces an impression upon my mind, and I stir, and slowly open my eyes. As yet I have really not seen anything. With my eyes open my mind still sleeps—but in a few seconds comes a possessing sense of well-being. Obeying some stimulus, not recognized by the senses as yet, I begin to stretch and yawn, then close my eyes and settle down into my pillows as for another nap. I am not aware that I am I, that I am awake, that I have yawned and stretched. I have a pleasant, half-dreamy feeling, but could not give it a name. For those few seconds this is all my world—a pleasant drowsiness, a being possessed by comfort. My consciousness is mere awareness—a pleasant awareness of uncomplicated existence. In another moment or two it is a consciousness of a day's work or pleasure ahead, the necessity of rising, dressing, planning the day, the alert reaction of pleasure or displeasure to what it is to bring, the effort to recall the dreams of sleep—the complicated consciousness of the mature man or woman. But I started the day with a mental condition close to pure sensation, a vague feeling of something different than what was just before.

Or this bare consciousness may come in the moment of acute shock, when the sense of suffering, quite disconnected from its cause, pervades my entire being; or at the second when I am first "coming back" after a faint, or at the first stepping out from an anesthetic. In these experiences most of us can recall a very simple mental content, and can prove to our own satisfaction that there is such a thing as mere awareness, a consciousness probably close akin to that of the lower levels of animal life, or to that of the newborn babe when he first opens his eyes to life.

Consciousness, then, in its elements, is the simplest mental reaction to what the senses bring.

How shall we determine when consciousness exists? What are its tests?

The response of the mind to stimuli, made evident by the body's reaction, gives the proof of consciousness in man or lower animal.

But what do we mean by a stimulus?

Light stimulates me to close my eyes when first entering its glare from a dark room, or to open them when it plays upon my eyelids as I sleep and the morning sun reaches me. It is a stimulus from without.

The fear-thought, which makes my body tremble, my pupils grow wide, and whitens my cheeks, is a stimulus from within.

An unexpected shot in the woods near-by, which changes the whole trend of my thinking and startles me into investigating its cause, is a stimulus from without causing a change within.

A stimulus, then, is anything within or without the body that arouses awareness; and this is usually evidenced by some physical change, however slight—perhaps only by dilated pupils or an expression of relief. When we see the reaction of the body to the stimulus we know there is consciousness. On the other hand, we cannot say that consciousness is always absent when the usual response does not occur; for there may be injury to organs accounting for the lack of visible reaction, while the mind itself may respond. But with due care, in even such cases, some external symptoms of response can usually be found if consciousness exists.

We have already realized how complex, intricate, and changing is fully developed consciousness.

The Unconscious

But the mind of man knows two distinct conditions of activity—the conscious and the unconscious. Mind is not always wide awake. We recognize what we call the conscious mind as the ruling force in our lives. But how many things I do without conscious attention; how often I find myself deep in an unexplainable mood; how the fragrance of a flower will sometimes turn the tide of a day for me and make me square my shoulders and go at my task with renewed vigor; or a casual glimpse of a face in the street turn my attention away from my errand and settle my mind into a brown study. Usually I am alert enough to control these errant reactions, but I am keenly aware of their demands upon my mind, and frequently it is only with conscious effort that I am kept upon my way unswerved by them, though not unmoved.

When we realize that nothing that has ever happened in our experience is forgotten; that nothing once in consciousness altogether drops out, but is stored away waiting to be used some day—waiting for a voice from the conscious world to recall it from oblivion—then we grasp the fact that the quality of present thought or reaction is largely determined by the sum of all past thinking and acting. Just as my body is the result of the heritage of many ancestors plus the food I give it and the use to which I subject it, so my mind's capacity is determined by my inheritance plus the mental food I give it, plus everything to which I have subjected it since the day I was born. For it forgets absolutely nothing.

"That is not true," you say, "for I have tried desperately to remember certain incidents, certain lessons learned—and they are gone. Moreover, I cannot remember what happened back there in my babyhood."

Ah, but you are mistaken, my friend. For you react to your task today differently because of the thing which you learned and have "forgotten." Your mind works differently because of what you disregarded then. "You" have forgotten it, but your brain-cells, your nerve-cells have not; and you are not quite the same person you would be without that forgotten experience, or that pressing stimulus, which you never consciously recognized, but allowed your subconsciousness to accept. Some night you have a strange, incomprehensible dream. You cannot find its source, but it is merely the re-enacting of some past sensation or experience of your own, fantastically arrayed. Some day you stop short in your hurried walk with a feeling of compulsion which you cannot resist. You know no reason for it, but some association with this particular spot, or some vague resemblance, haunts you. You cannot "place" it. One day you hit the tennis-ball at a little different angle than you planned because a queer thought came unbidden and directed your attention aside. Again, under terrific stress, with sick body and aching nerves, you go on and do your stint almost mechanically. You do not know where the strength or the skill is derived. But your unconscious or subconscious—as you will—has asserted itself, has usurped the place of the sick conscious, and enabled you automatically to go on. For we react to the storehouse of the unconscious even as we do to the conscious.

Remember that the unconscious is simply the latent conscious—what once was conscious and may be again, but is now buried out of sight.

The mind may be likened to a great sea upon which there are visible a few islands. The islands represent the conscious thoughts—that consciousness we use to calculate, to map out our plans, to form our judgments. This is the mind that for centuries was accepted as all the mind. But we know that the islands are merely the tops of huge mountain-ranges formed by the floor of the sea in mighty, permanent upheaval; that as this sea-floor rises high above its customary level and thrusts its bulk above the waters into the atmosphere, is the island possible.

Just so there can be no consciousness except as that which is already in the mind—the vast subconscious material of all experience—rises into view and relates itself through the senses to an outside world. We speak very glibly of motion, of force, of power. We say "The car is moving now." But how do we know? Away back there in our babyhood there were some things that always remained in the same place, while others changed position. The changing gave our baby minds a queer sensation; it made a definite impression; and sometimes we heard people say "move," when that impression came. Finally, we call the feeling of that change "move," or "movement," or "motion." The word thereafter always brings to our minds a picture of a change from one place to another. The process—the slow comprehending of the baby mind—was buried in forgetfulness even at the time. But had not the subconscious been imprinted with the incident and all its succeeding associations, that particular phenomenon we could not name today. It would be an entirely unique experience. So our recognition of the impression is merely the rising into consciousness of the subconscious material in response to a stimulus from the outside world which appeals through the sense of sight. We can get no response whatever except as the stimulus asking our attention is related by "like" or "not like" something already experienced; that is, it must bear some relation to the known—and perhaps forgotten—just as the island cannot be, except as, from far down below, the sea-floor leaves its bed and raises itself through the deeps. The visible island is but a symbol of

12

the submarine mountain. The present mental impression is but proof of a great bulk of past experiences.

And so we might carry on the figure and compare the birth of consciousness to the instant of appearance of the mountain top above the water's surface. It is not a new bit of land. It is only emerging into a new world.

"But," you ask, "do you mean to assert that the baby's mind is a finished product at birth; that coming into life is simply the last stage of its growth? How unconvincing your theory is."

No, we only now have the soil for consciousness. The island and the submarine mountain are different things. The sea-floor is transformed when it enters into the new element. An entirely different vegetation takes place on this visible island than took place on the floor of the sea before it emerged. But the only new elements added to the hitherto submerged land come from the new atmosphere, and the sea-floor immediately begins to become a very different thing. Nevertheless, what it is as an island is now, and forever will be due, primarily, to its structure as a submarine mountain. In the new atmosphere the soil is changed, new chemical elements enter in, seeds are brought to it by the four winds—and it is changed. But it is still the sea-floor transformed.

Just so the baby brain, complete in parts and mechanism at birth, is a different brain with every day of growth in its new environment, with every contact with the external world. But it is, primarily and in its elements, the brain evolved through thousands of centuries of pushing up to man's level through the sea of animal life, and hundreds of centuries more of the development of man's brain to its present complete mechanism through experience with constantly changing environment.

Hence, when the baby sees light and responds by tightly shutting his eyes, then later by opening them to investigate, his sensation is what it is because through the aëons of the past man has established a certain relation to light through experiencing it. To go further than this, and to find the very beginning, how the first created life came to respond to environment at all, is to go beyond the realm of the actually known. But that he did once first experience his environment, and establish a reaction that is now racial, we know.

So our baby soon shows certain "instinctive" reactions. He reaches out to grasp. He sucks, he cries, he looks at light and bright objects in preference to dark, he is carrying out the history of his race, but is making it personal. He has evolved a new life, but all his ancestors make its foundation. The personal element, added to his heritage, has made him different from any and all of his forebears. But he can have no consciousness except as a bit from the vast inherited accumulation of the past of his ancestors, of all the race, steps forth to meet a new environment.

And again you ask, "How came the first consciousness?"

And again I answer, "It is as far back as the first created or evolved organism which could respond in any way to a material world; and only metaphysics and the God behind metaphysics can say."

We only know that careful laboratory work in psychology—experiments on the unconscious—today prove that our conscious life is what it is, because of: first, what is stored away in the unconscious (i. e., what all our past life and the past life of the race has put there); second, because of what we have accepted from our environment; and this comprises our material, intellectual, social, and spiritual environment.

Consciousness is Complex

The one fact we want at this stage of our inquiry is simply this: that consciousness, awaking at birth, very soon becomes complex. However single and simple in content immediate consciousness may be, it is so intimately linked with all preceding experience that a pure sensation is probably never known after the first second of life. As the sensation is registered it becomes a basis for comparison. That first sensation, perhaps, was just a feeling of something. The next is a feeling of something that is the same, or is not the same, as the first. So immediately perception is established. The baby consciousness recognizes that the vague feeling is, or is not, that same thing. And from perception to a complex consciousness of perceptions, of ideas, of memories and relations, and judgments, is so short a step that we cannot use our measuring rods to span it.

Thus through the various stages of life, from infancy to maturity, the conscious is passing into the unconscious, only to help form later a new conscious thought. Hence the conscious thought is determined by the great mass of the unconscious, plus the external world.

But every thought, relegated to the unconscious, through its association there—for it is plastic by nature—comes back to consciousness never quite the same, and meets never quite the same stimulus. And as a result a repeated mental experience is never twice exactly the same. So the conscious becomes the unconscious and the unconscious the conscious, and neither can be without the other.

Our problem is to understand the workings of the mind as it exists today, and to try to find some of its most constructive uses; and on that we shall focus attention. To that end we must first examine the various ways in which consciousness expresses itself.

We have recognized two distinct mental states—the conscious and the unconscious—and have found them constantly pressing each on the other's domain. Our study of consciousness reveals the normal in the aspects of sleeping and waking, also various abnormal states. Consciousness may become excited, depressed, confused, delirious, or insane. We shall consider later some of the mental workings that account for these abnormal expressions. At present let us examine the mind's activities in sleep and in delirium.

Consciousness in Sleep

Sleep seldom, if ever, is a condition of utter unconsciousness. We so frequently have at least a vague recollection, when we wake, of dreaming—whether or not we remember the dream material—that we are inclined to accept sleep as always a state of some kind of mental activity, though waking so often wipes the slate clean. A new word which serves our purpose well has come into common use these last years, and we describe sleep as a state of rest of the conscious mind made possible as weariness overpowers the censor, and this guard at the gate naps. The censor is merely that mental activity which forces the mind to keen, alert, constructive attention during our waking hours, a guard who censors whatever enters the conscious mind and compares it with reality, forcing back all that is not of immediate use, or that is undesirable, or that contradicts established modes of life or thought. In sleep we might say that the censor, wearied by long vigilance, presses all the material—constantly surging from the unconscious into consciousness, there to meet and establish relations with matter—back into the unconscious realms, and locks the door, and lies and slumbers. Then the half- thoughts, the disregarded material, the unfit, the unexpressed longings or fears, the forbidden thoughts; in fact, the whole accumulation of the disregarded or forgotten, good, bad, and indifferent—for the unconscious has no moral sense—seize their opportunity. The guard has refused to let them pass. He is now asleep. And the more insistent of them pick the lock and slip by, masquerading in false characters, and flit about the realms of the sleeping consciousness as ghosts in the shelter of darkness. If the guard half-wakes he sleepily sees only legitimate forms; for the dreams are well disguised. His waking makes them scurry back, sometimes leaving no trace of their lawless wanderings. So the unconscious thoughts of the day have become sleep-consciousness by play acting.

Consciousness in Delirium

At this time of our study it will suffice to say that in delirium and in insanity, which we might very broadly call a prolonged delirium, the toxic brain becomes a house in disorder. The censor is sick, and sequence and coherence are lost as the thronging thoughts of the unconscious mind press beyond the portals into consciousness, disordered and confused. We shall later find, however, that this very disorder falls into a sort of order of its own, and a dominant emotion of pain or ecstasy, of depression or fear, of exaltation or depreciation calls steadily upon the stored away incidents and remembered, related feelings of the past and interprets them as present reality. The censor of the sick brain is stupefied by toxins, shock, or exhaustion, and the citadel he is supposed to guard is thronged with besiegers from every side. The strongest—i. e., those equipped with most associations pertinent to the emotional status at the time—win out, occupy the brain by force, and demand recognition and expression from all the senses, deluding them by their guise of the reality of external matter.

We find consciousness, then, determined by all past experience, by an external world, and by its organ of expression—the brain.

Consequently, our psychology leads us into anatomy and physiology, which, probably, we have already fairly mastered. In rapid review, only, in the following chapter we shall consider the organs of man's consciousness, the brain, spinal cord, and the senses, and try to establish some relation between the material body and its mighty propelling force—the mind.

CHAPTER III
ORGANS OF CONSCIOUSNESS

Nothing is known to us until it has been transmitted to the mind by the senses. The nerves of special sense, of sight, hearing, smell, taste, touch, the temperature sense ("hot or cold" sense), the muscular sense (sense of weight and position), these, and the nerves controlling voluntary motion, form the peripheral, or surface, nervous system. This acts as a connecting medium between the outside world and the central nervous system, which is composed of the brain and spinal cord. We might liken the nerves, singly, to wires, and all of them together to a system of wires. The things of the external world tap at the switchboard by using the organs of special sense; the nerves, acting as wires, transmit their messages; at the switchboard is the operator—consciousness—accepting and interpreting the jangle of calls.

The recognition by the brain of the appeals coming by way of the transmitting sense, and its interpretation of these appeals, is the mind's function of consciousness, whether expressed by thinking, feeling, or willing.

The Central and Peripheral Nervous Systems in Action

I am passing the open door of a bake-shop, and a pervading odor fills the air. I think "hot rolls," because my organ of smell—the nose—has received a stimulus which it transmits along my olfactory nerves to the brain; and there the odor is given a name—"hot rolls." The recognition of the stimulus as an odor and of that odor as "hot rolls" is consciousness in the form of thinking. But the odor arouses desire to eat—hunger; and this is consciousness in the form of feeling. The something which makes me walk into the shop and buy the rolls is consciousness in the form of willing. The sensory appeal from the outside world gained admission through the sense of smell; this transmitted the message, and consciousness recognized the stimulus, which immediately appealed to my hunger and incited action to satisfy that hunger.

The ear of the operator in the telegraph office, again, might illustrate consciousness. It must be able to interpret mere clickings into terms of sense. To the operator the sounds say words, and the words are the expression of the object at the other end of the wire. The brain is the receiving operator for all the senses, which bring their messages in code, and which it interprets first as sound, vision, taste, touch, feel, smell, temperature; then more accurately as words, trees, sweet, soft, round, acrid, hot.

The mind can know nothing except as the stimulus is transmitted by sense-channels over the nerves of sense, and received by a conscious brain. A baby born without sight, hearing, taste, smell, or touch would remain a mere bit of clay. He could have no awareness. But so long as any one sense channel remains open the mind may acquire some knowledge.

Suppose I am paralyzed, blind, and deaf, and you put a tennis-ball into my hand. I cannot tell you what it is, not even what it is like. It means nothing whatever to me, for the sense channels of touch, sight, and hearing, through which alone it could be impressed upon my brain, are gone. Suppose I am blind and deaf, but have my sense of touch intact; that I never saw or touched or heard of a tennis-ball before, but I know "apple" and "orange." I can judge that the object is round, that it is about the size of a small orange or apple. It is very light, and has a feel of cloth. I know it to be something new in my experience. You tell me in the language of touch that it is "tennis-ball"; and thereafter I recognize it by its combination of size, feel, and weight, and can soon name it as quickly as you, who see it.

Suppose I am blind and my hands are paralyzed, but I have my hearing. You tell me this is a tennis-ball, and if I have known "tennis-ball" in the past, I can describe it to you. It has been impressed upon my brain through my sense of hearing; and memory im mediately supplies the qualities that go with "tennis-ball."

But if none of the senses has ever developed, my brain can receive no impression whatever; it cannot have even the stimulus of memory. Hence conscious mind cannot be, except as some sense-channel or channels have been opened to carry thought material to the brain. So far as we know today, in this world, mind is absolutely dependent upon the sense organs and the brain—upon matter—for existence.

The Sympathetic Nervous System

Associated with the central nervous system by connecting nerves—but located outside of it in various parts of the body—are groups of nerve-cells (gray matter) and their fibers, forming what we call the sympathetic nervous system—the direct connecting link between mind and body.

The central nervous system is the director of all conscious action of the body; the sympathetic orders all unconscious action.

The beating of the heart, the contraction of the blood-vessels, hence the flowing of the blood, the processes of digestion, the functioning of the glands, are all directed by the sympathetic. In other words, the central nervous system normally controls the movements of the voluntary muscles; the sympathetic controls those of the involuntary muscles.

The quick blush, the sudden paling of the cheeks, the start of fear, the dilated pupils of fright are the direct result of the action of involuntary muscles under control of the sympathetic system. The stimulus is received by the central nervous system; the fibers connecting the central and the sympathetic systems carry the message quickly to the latter, which immediately respond by ordering contraction or expansion of involuntary muscles. So tears flow, we breathe freely again or we quake and tremble, our pupils widen or contract, the heart beats suffocatingly, or seems almost to stop.

The sympathetic system, as the name implies, is influenced by suggestions from the emotions rather than from the intellect. We might say that it is controlled by the "feeling mind" rather than the thinking mind, for intellect cannot influence it in the least.

The wise nurse, who knows something of the laws of the mind, soon realizes that the sympathetic nervous system, rather than physical disability, causes many indigestions, headaches, diarrheas, dry mouths, chills; is responsible for much nausea, much "exhaustion," etc. When she has had wider experience she finds that almost any known physical disorder can be unconsciously imitated by the suggestible patient, whose sympathetic nervous system causes physical reactions to respond to the feelings of a sick mind. Let the nurse remember, however, that is it not for her to decide whether the disorders from which her patient suffers are of physical or nervous origin. It is for her, on the other hand, to study her patient's mentality and reactions, and to become expert in reporting symptoms of nervous as well as of physical significance.

CHAPTER IV
RELATION OF MIND AND BODY

We have found that mind is entirely dependent upon the bodily organs for its existence. Is the body in the same way dependent upon the mind? Can the mind die and the body go on?

Given a perfect body with unblocked sense channels, and put the mind to sleep, paralyze the central nervous system with alcohol in sufficient quantity so that the undamaged peripheral nervous system—the senses—can obtain no response or recognition from it, and that perfect body is as useless for the time as if dead. But here comes proof of the remarkable hold of the body on life. The unconscious mind takes up the burden of directing the sympathetic nerves to stimulate the muscles of breathing. The unconscious sees to the beating of the heart. It directs the contraction of the blood-carrying vessels. It maintains certain vital processes of secretion. Thus automatically life goes on; the body still reacts to a limited field of stimuli, and consciousness recognizes it not. But when the unconscious mind ceases to function, then, indeed, does the body die. Yet the conscious mind may "die" and the body live on, so long as the unconscious continues its activity.

It is possible for the human body to live for years, utterly paralyzed, with many of the senses gone, with no consciousness of being—if cared for by other persons—a merely vegetable existence. The current of power is broken; but the spark is still glowing, though utterly useless because connected with nothing. And it may continue to glow for some time while properly stimulated from outside sources.

We might liken the mind to the boiler in which steam is generated, and the body to the engine which the steam runs. If the boiler bursts, the engine stops; but it may not be otherwise damaged. It simply cannot carry out its main function of motion any longer. The fires under the boiler are still burning and can be kept burning so long as fuel is provided, but the connection is broken and the great bulk of iron is a useless thing in that it can no longer fulfil its purpose.

In just such a way may the mind be paralyzed; but the spark of life, which has through all the years kindled the now lost mind to action, may still remain—a useless thing, which would die away if not tended from without by other bodies whose minds are still intact.

But in the demented mind consciousness still remains, the awareness of the young child or baby stage of life. The connection between the upper or conscious brain centers and the body has been tampered with; it no longer is direct, but breaks off into switch-lines. But the contact still holds between the lower or unconscious mind and the body; so the automatic body functions go on, directed as they were in babyhood before the independent mind assumed control. Hence, when all acute consciousness is finally gone, the unconscious mind, a perfect automaton, may still carry out the simplest vegetative activities of existence.

When body is dead, mind, so far as its reactions to the world we know are concerned, ceases to act. But when the conscious mind is "dead" the body may yet live as a vegetable lives, with all its distinctively human functions lost. Motionless, save for the beating of the heart and the reaction of the lungs to air, the body may still be alive, though the mind long since has ceased all earthly activity.

So we discover that an organ of mind is an essential, here, to life of mind, and that mind only can induce this organ to any action above the vegetative stage. But, on the other hand, we find that life can exist without conscious mind, even if untended by others, for a limited time.

If the direct nerve connections between the brain and the hand, the brain and the foot, or the brain and the trunk are cut off, the mind henceforth realizes nothing of that part except as the sense of sight reports upon it; for the optic nerves relate the hand and mind, through this sense, as truly as the motor nerves which carry the mind's message for motion to the hand, and the sensory nerves which carry back to the mind the hand's pain. But let the optic nerve be inert, the sensory and motor connections broken between brain and hand, or foot and trunk, or brain and trunk, and the hand or foot may be amputated and the mind never sense the fact; the trunk may be severely injured and the mind be serenely unconscious. So the brain in man is "the one immediate bodily condition of the mental operations." Take away all the brain and man's body is a useless mass of protoplasm.

The brain's varied and intricate nerve connections with all parts of the body, through nerves branching from the main trunks in the spinal cord, we shall not discuss, for you know them through your study of anatomy. For the purpose of our psychology we need consider only two of the main divisions of the brain—the cerebrum, which includes what we call the right and left hemispheres, and the cerebellum.

The Cerebrum or Forebrain

For convenience the various lobes of the cerebrum are known as frontal, temporal, parietal, and occipital, according to the parts of the brain referred to: as forehead, temples, crown, or occiput. The cerebellum, or hind brain, is also divided into two hemispheres, and is situated behind and below the hemispheres of the cerebrum.

A system of localization has been roughly mapped out, the result of careful laboratory work on animals and of studying the loss of various functions in human beings as related to the location of brain injuries.

From these experiments it seems proved that consciousness belongs only to the cortex or surface of the upper brain, and that the vast realm of the unconscious belongs to the lower brain centers. Hence the cortex is the organ of consciousness, and the lower centers are the repository of the unconscious until it again becomes conscious.

The motor zone of the cortex we now know to be situated in the convolutions bordering the fissure of Rolando. Vision is evidently excited from the occipital lobes, though not yet conclusively proved. Smell, presumably, is located in the temporal lobes. Considered action is directed from the upper hemispheres only. It is significant that the hemispheres of the cerebrum are also accepted as the seat of memory for man—that intellectual quality which makes him capable of acting from absent stimuli, stimuli only present to memory; which makes it possible for him to reason the present from the experiences of the past.

But in all animal life, except the higher forms, the control of action is from the lower brain centers, centers which respond only to present objects. With them memory, as man knows it, is lacking; but the reactions of the past are indelibly imprinted upon motor nerves and muscles, so that when the present object presses the button, as it were, calling forth the experience of the race, the animal instinctively reacts.

But of what use to man, then, are the lower brain centers?

In man, as in lower animals, they care for the vegetative functions of life, so that our blood continues to circulate, the air enters and leaves our lungs, digestion is carried on, with no assistance from the upper centers, the hemispheres of the cerebrum being thus left free for concentration on the external world of matter, which it can transform into a world of thought.

It is the lower or vegetative brain that may still exist and keep life intact when the functions of the cerebrum are destroyed. We can say, then, of the brain as a whole that it is the organ of the mind, the sine qua non of the mind, the apparatus for the registration of sense impressions. The senses themselves are the rudiments of mind, are the means by which stimuli alighting on sense organs enter consciousness; for the nerves of special sense immediately carry the impetus to the brain, where it is recognized as the "not me," the something definitely affecting the me, and demanding reaction from the me.

The functions of the cerebrum we find grouping themselves in three classes: intellect, emotion, and volition, more simply, thinking, feeling, and willing; and we find no mental activity of the normal or abnormal mind which will not fall into one of these groupings. This does not mean that one part of the brain thinks, another part wills, another part feels; for in the performance of any one of these functions the mind acts as a whole. Our thinking or our willing may be permeated with feeling, but the entire mind is simply reacting simultaneously upon various stimuli.

CHAPTER V
THE NORMAL MIND

Mind, we found, is born in the form of consciousness when the outside world impresses itself upon the brain-cells by way of the senses. This consciousness, observation and experiment prove, is first a feeling one, later a feeling-thinking-willing one. The mind, then, is really the activity of the brain as it feels, as it thinks, as it wills. We express this in descriptive terms when we speak of mind as the flow of consciousness, the sum of all mental associations, conscious and unconscious. For mind is never a final thing. Looking within at our own mental processes we find that always our thought is just becoming something else. We reach a conclusion, but it is not a resting place, only a starting place for another. My thought was that a moment ago, but while it was that it was becoming this, and even now it is becoming something else.

Thinking is mind. Feeling is mind. Willing is mind. But for the sake of clearness we speak of feeling, thinking, and willing as being functions of mind. Mind acts by using these powers. But to what end does it act? What purpose does it serve? For these functions are not the reasons of being for the mind, even as motion — while the immediate purpose of the locomotive — is not its chief end. The steam engine may stand in the same spot while its wheels revolve madly; it may move along the tracks alone, and accomplish nothing; or it may transport a great train of loaded cars. Unless it moves to some definite point and carries merchandise or people there, it is a useless, indeed, a dangerous invention. We find, in fact, that it functions to the very definite end of taking man and his chattels to specified places.

And so it is with the mind. If it is thinking and feeling and willing only for the sake of exercising these mental powers, it might better not be. But what end do we actually find these functions serving?

Mind, with its powers of thinking, feeling, and willing, gives an external world of matter; an internal world of thought, and so relates them to each other as to make them serve man's purposes. Thus these functions exist for accomplishment.

In the solving of a problem, for instance, the mind thinks, primarily; in the enjoyment of music it feels, primarily, though its feeling may be determined by the intellectual verdict on the music; in forcing its owner to sit at the piano and practice in the face of strong desire to attend the theater, it wills, primarily. Now one of its functions predominates; now another. But the whole mind, not a feeling section, or a thinking section, or a willing section, operates together to produce action. When I play the piano it calls on all my mind. I think the music. I feel it. I make my fingers play it. But the thinking, the feeling, and the willing act together to result in the fingers playing.

The mind, then, is an instrument of achievement. It fulfils its purpose when it makes matter serve useful ends.

Emotion or feeling is the function of the mind which associates a sense of pleasure or pain with every thought or act.

Feeling is the affective state of mind. By this we mean that it has the power to move us. And this emotion primarily does; for our feeling of pleasure or pain moves us to action, as well as precedes and accompanies and follows action. The word emotion is usually employed to denote an acute feeling state, while the word mood denotes a prolonged feeling condition, i. e., a less acute emotional state. The word feeling, however, is used to cover both; for in each case the sensational element manifests itself in a definite physical affect, pleasurable or painful in some degree.

Thinking is a conscious mental activity exercised to evolve ideas from perceptions, and to combine and compare these ideas to form judgments.

Intellection, or thinking, might be explained as the mental process which converts sensation into percepts, groups percepts to form concepts or ideas, stores away ideas and sensations for future use, and recalls them when needed—the recalling being memory—and by reason combines, compares, and associates ideas to form judgments, then compares judgments to form new judgments. The process of intellect we name by terms denoting activity, such as intellection, thinking, the stream of thought, and the latter describes it most truly.

Volition or will is the function of the mind which compels the expression of thought or feeling in action.

For clarity we might indicate the mind and its functions in the following diagram:

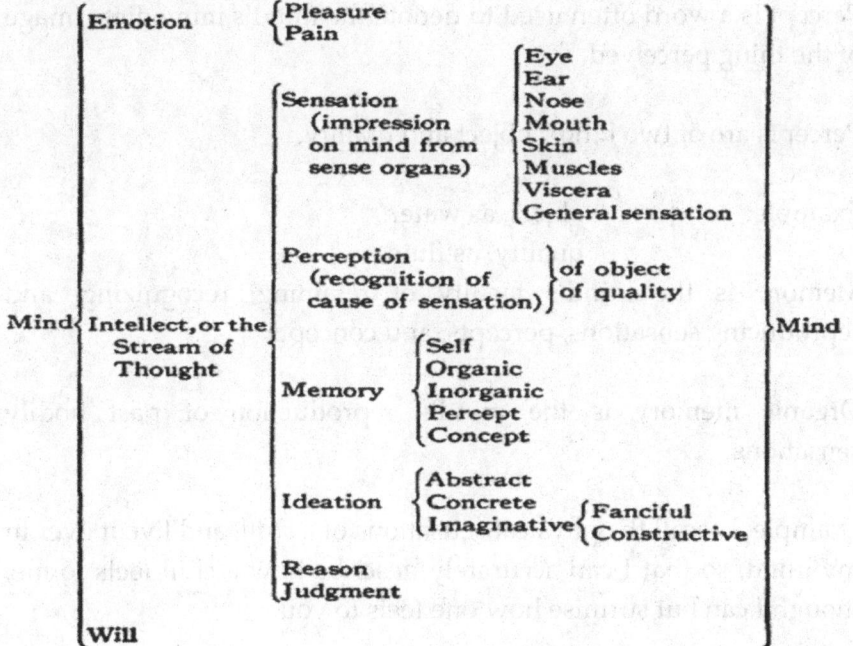

The following terms are ones constantly used in psychology, and are briefly defined that there may be no haziness in their application.

Sensation is the uninterpreted response of the mind to stimuli brought by sense organs.

Examples: Feeling of { hot.
cold.
pain.

Sensation may arouse instinct and cause reflex action, or start a feeling state, or a train of thought.

Perception is the conscious recognition of the cause of a given sensation.

Example: { fluid—water.
cold—snow.
pain—cut.

Percept is a word often used to denote the mind's immediate image of the thing perceived.

Percepts are of two kinds: object and quality.

Example: { object, as water.
quality, as fluid.

Memory is the mind's faculty of retaining, recognizing, and reproducing sensations, percepts, and concepts.

Organic memory is the mind's reproduction of past bodily sensations.

Example: I recall the physical sensations of a chill, and live it over in my mind, so that I can accurately describe how a chill feels to me, though I can but surmise how one feels to you.

Inorganic memory is the mind's reproduction of its own reactions in the past.

Example: Myself having a chill, how I acted; what I thought and my emotions during that chill.

Ideation is the mind's grouping of percepts by the aid of memory, to form concepts.

Example: I perceive color, form, mouth, eyes, nose, chin, etc. These percepts I combine as a result of past experience (memory) to form my concept, face; and the process of combining is ideation.

Concepts are mental representations of things or qualities, i. e., of object or quality percepts.

We might say that the percept is the mind's immediate image of a thing or quality, and the concept is the result of the storing up and grouping and recombining of percepts. Thus a lasting mental picture is secured; and my idea of horse, for instance, is so clear and definite a thing in my mind that if I should never again see a particular horse, I should yet always be able to think accurately of a horse.

Concepts are of two kinds—concrete and abstract.

A concrete concept, or concrete idea (for concept and idea are interchangeably used), is an idea of a particular object or quality.

Examples: This wine-sap apple (object concept).
 This sweet orange (quality concept).

An abstract concept, or abstract idea, is a mental reproduction of a quality or an object dissociated from any particular setting or particular experience.

Abstract ideas are of two kinds. We speak of them as abstract object concepts and as abstract quality concepts. An abstract object concept we might call a generalized idea, an idea comprehending all objects having certain things in common.

Example: My idea of animal includes many scores of very different individual animals, but they all have bodies and heads and extremities. They all have some kind of digestive apparatus; they breathe, and can move.

An abstract quality concept is easier to think than to explain. It is as though the mind in considering a multitude of different objects found a certain quality common to many of them, and it "abstracted," i. e., drew this particular quality, and only this, from them all, and then imagined it as a something in itself which it calls redness, or whiteness, or goodness. Thereafter, whenever it finds something like it anywhere else again it says, "That is like my redness." So I call it "red." In other words, consciousness thereafter can determine in a newly discovered object something it knows well merely because that something corresponds to a representation which experience and memory have already formed.

These comprehensive concepts, or universals, as some psychologists term them, the mind, having pieced together from experience and memory, holds as independent realities, not primarily belonging to this or that, but lending themselves to this or that. For example: My mind says "white," and sees white in some object. But I see the white only because my mind has a quality concept, whiteness. This outside object corresponds to my concept. I recognize the likeness and call it "white."

I speak of goodness, or purity, of benevolence; or of fulness, emptiness, scantiness. There is no object or quality in the outside world I can say is goodness, or fulness. But I do see things in the external world through my ideas of goodness or fulness that correspond to these ideas. They have some of the qualities the ideas embrace; and so I point them out and say, "This represents purity;

that, impurity"; or, "This is full, that is empty." One satisfies my concept of purity, while the other does not. One fulfils my concept of fulness; the other does not. And because we can never point out any one quality in the outside world and say "This is purity, and all of purity; this is goodness; or this good plus this good plus this makes all of goodness"; because of this impossibility we speak of these concepts as having reality somewhere. They are absolutes, universals, abstract quality concepts—the unfound all of which the things we call pure and good are but the part.

Apperception is the process of comparing the new with all that is in the mind, and of classifying it by its likeness to something already there.

With an abstract idea of an object in mind we very deftly, through the use of memory and constructive imagination, deduce the whole from the part recognized as familiar.

Example: In walking through the field, along the bank of the brook, I glimpse under the low-hanging branches of the weeping willow a restlessly moving hoof. I see a certain kind of hoof and only that. Or I hear a lowing sound. And I say "cow." I have not seen a cow, but only a part which tells me a cow is there; for all the cows I ever saw had hoofs of that general description, and so it fits into my concept cow, and into no others. Or I have heard cows, only, give that lowing sound before. From my perception, then, of hoof or sound I apperceive cow. Memory relates that hoof or that lowing sound to a certain kind of animal known in the past; and constructive imagination draws in all the rest of the picture that belongs with it.

Again, we may apperceive an object or quality from our recognition of something which in our experience has been associated, under those particular circumstances, with only that object or quality. I see smoke on the ocean's far horizon, and I decide instantly, "a steamer." I have not perceived any steamer, but only something that "goes with it," as it were. I see the ship with my mind, not with my eyes; for I know that a cloud of smoke out there always has, in my past experience, represented just that. I compare the newly appearing stimulus—smoke in that particular location—with all that is associated with it in my mind, and classify it with the known. I apperceive "steamer."

In apperception, then, we construct from the known actually perceived by the senses, the unknown. How does the child realize that the moving speck on the distant hillside is his father? There is nothing to indicate it except that it is black and moves in this direction. But experience tells Johnny that father comes home that way just about this time. Moreover, it says that father looks so when at that distance. When Johnny is as sure it is his father as if he could see his face close beside him he has apperceived him. The speck on the hill is the newly arriving stimulus. Johnny compares it with what corresponds to it in his mind's experience and proclaims, as a fact, that he sees his father.

Reason is the mind's comparison and grouping of concepts to form judgments, and its association of judgments to form new judgments.

Example: My concept man includes the eventual certainty of his death. My concept mortal means "subject to death." Therefore my judgment is, "Man is mortal." Reason has compared the concepts and found that the second includes the first.

Judgment is the mind's decision arrived at through comparing concepts or other judgments.

Example: Man is mortal is my decision after comparing the concepts man and mortal and finding that the latter really includes the former. Judgment at the same time says that "Mortals are men," is not a true conclusion. For in this case the first concept is not all included in the second. Mortals are all life that is subject to death.

We may assume personal consciousness even as we recognize an individual body. Psychology does not deal with any awareness separated from a person. It knows no central mind of which you partake or I partake, and which is the same for us both. A universal consciousness would simply mean one which is the sum of yours and mine and everybody's who lives today, or who has ever lived. So by personal consciousness the psychologist means his consciousness, or yours, or mine. But they can never be the same; for mine is determined by my entire past and by how things and facts and qualities affect me; and yours, by your past, and by things and facts and qualities, and by how they affect you.

Personal consciousness is the mind's recognition of self; and as the self changes with every added experience, so personal consciousness is modified.

Stream of thought is a term James has brought into common usage to illustrate the fact, already stressed, that thinking, as we know it, is never static, is never one thing, one percept, one concept, one judgment; but is a lot of these all together, just beginning to be or just beginning to change into something else. We never know a concept, for instance, except as it is a part of our entire consciousness, related to all the rest; just as we do not know the drop of water in the brook as it flows with the stream. We can take up one on our finger-tips, however, and separate it from all the rest. But analyzed in the laboratory, this drop will contain all the elements that a pint or gallon or a barrel of the same water contains. The drop is what it is because the stream has a certain composition. We only have a brook as drops of rain combine to make it, but we also have only the drops as we separate them from the steam.

Imagination is the combining by the mind, in a new way, things already known.

This may be either into fantastic groupings divorced from reality, or into new, possible, rational groupings not yet experienced. So imagination is of two kinds, the fantastic and the constructive. Fantastic imagination, or fantasy, gives us gnomes, fairies, giants, and flying horses, and all the delights of fairy tales. Constructive imagination is the basis for invention, for literature, and the arts and sciences.

The word thinking, defined early in this chapter, is broadly used to denote the sum of all the intellectual faculties. Thinking is really the stream of thought.

CHAPTER VI
THE NORMAL MIND (Continued)

Instinct

We have found that the mind's chief end is action, of itself, or of its body. But what are its incentives to action?

We see the very young baby giving evidences of an emotional life, living in an affective, or feeling environment, leading a pleasure-pain existence, from the first. He acts as desire indicates. But from the very moment of his birth he performs actions with which he cannot as yet have a sense-memory connection, because he is doing them for the first time. How can he know how to respond to stimuli from the very beginning?

No other possible explanation offers itself than that he is born with certain tendencies to definite action. These we call instincts—man's provision to keep him going, as it were, till reason develops. Instincts are handed down from all the past. Definite tendencies, they are, to certain specific reflex actions in response to certain sensations. These responses, from the very beginning of animal life, have been toward avoiding pain, and toward receiving pleasure. It is as though the stimulus presses the trigger—instinct—and the muscle responds instantly with reflex action. This mechanism is the means of protection and advancement, and takes largely the place of intelligence in all animal life. It is what makes the baby suck and cry, clutch and pull, until a sense memory is established. So instinct is really race memory. We call instinctive those immediate, unthought reactions which are the same with all mankind.

The pugnacious instinct—the desire to fight—is the natural reaction of every human being of sane mind to attack. The inner necessity of avenging is so strong in the child or man of untrained mind or soul that he acts before he thinks. He strikes back, or shoots, or plots against his enemies. Only rare development of spirit or the cautious warning of reason which foresees ill consequences, or a will trained to force control, can later make the instinct inactive.

Where instinct ends and sense memory, imitation, and desire step in is difficult to determine. Later in life probably most of what we consider instinctive action is simply so-called reflex action, depending on sense memory, action learned so young that it is difficult to distinguish it from the true reflex action, which is due only to race memory.

James, in his Talk to Teachers, gives us a partial list of the instincts. Thus:

Fear	Ownership	Shyness
Love	Constructiveness	Secretiveness
Curiosity	Love of approbation	The ambitious impulses:

- Imitation,
- Emulation,
- Pride,
- Ambition,
- Pugnacity

To this partial list we would add self-preservation, reproduction, etc.

But instincts conflict with each other, and man carries about with him in babyhood many of them which may have been very useful to his prehistoric ancestors, but which only complicate things for him. Fear and curiosity urge opposite lines of conduct. Love of approbation and shyness are opposed. Love and pugnacity are apt to be at odds. So, gradually, as intelligence increases, the child refuses to allow such impulses to lead him to action. When fear-instinct and love-instinct are at war, reason is provided to come to the rescue.

Instincts are racial tendencies of sensational or emotional states to determine action.

Instincts are the germs of habit, and when instinct would give rise to a reaction no longer useful, reason, abetted by new habit formation, in the normal mind, weakens instinct's force; and the habit is discarded and the instinct gradually declines.

In prehistoric times when food was scarce, and man had not learned the art of tilling the soil, hunger forced him to fight for what he got to eat. As there was often not enough to go around, he maimed or killed his fellow-man that he might have all he wanted, obeying the instinct to survive. So, now, the baby instinctively clutches for all that appeals to him. But an abundance of food for all, or the intelligent realization that co-operation brings more to the individual than does fighting, and a developed sense of responsibility toward others; or merely the fear of the scorn of fellow beings, or the desire to be protected by the love of his kind; perhaps a genuine love of people, acquired by spiritual development, puts the primitive habit of food-grabbing into the discard. Finally, the very instinct of self-preservation may be transformed into desire to serve others. No better illustration of this can ever be offered than the sacrifices of the World War.

Memory

No mind retains consciously everything that has ever impressed it. It is necessary that it put aside what ceases to be of importance or value and make way for new impressions. We found early in our study that the subconscious never forgets, but harbors the apparently forgotten throughout the years, allowing it to modify our thinking, our reactions. But the conscious mind cannot be cluttered with the things of little importance when the more essential is clamoring. So there is a forgetting that is very normal.

We forget numberless incidents of our childhood and youth; we may forget the details of much that we have learned to do automatically; but the subconscious mind is attending to them for us.

Do you know how to skate? and if so, do you remember just how you did it the first time? Probably all you recall is that you fell again and again because your feet would slip away from where you meant them to be. When you glide over the ice now it is as natural as walking, and as easy. You cannot remember in detail at all how you first "struck out," nor the position of your feet and arms and legs, which you felt forced to assume. At the time there was very real difficulty with every stroke—each one was an accomplishment to be attempted circumspectly, in a certain definite way. All you remember now is, vaguely, a tumble or two, soreness, and lots of fun.

We forget details we have intrusted to others as not a part of our responsibility. We forget the things which in no way concern us, in which we have no interest and about which we have no curiosity. And it is well that we do so. If it were not for the ability to forget, our minds would be like a room in which we have lived a lifetime, where we have left everything that has been brought into it since our birth. It would be piled ceiling high, with no room for us, and with difficulty only could we find what we want. As we grow from babyhood to childhood, from childhood to youth, from youth to maturity the room changes with us. We put off childish things. They are stored away somewhere, in an attic or basement, or destroyed. And day after day something new is added, displacing something else. In the case of the mind all these things are stored and cataloged in the subconscious, and forgotten, until some need causes us to look into our catalog-index and see the experience again, or some association calls it back, relating it to something new. So our discussion of the subconscious involved also a discussion of memory.

But what of the things we must use frequently and cannot find in our minds? What of absent-mindedness and faulty memory? In such cases our minds might be compared to a cluttered room full of things we need and want to use every day, but in confusion. We know where many of them are, the ones we care most about; but we have to rummage wildly to find the rest. We have no proper system of arrangement of our belongings. You laid down that book somewhere, absent-mindedly, and now you cannot tell where. You were thinking of something else at the time, and inattention proves a most common cause of poor memory. Perhaps you simply have more books than the room can hold in an orderly way, and so you crowded that one in some corner, and now have no recollection of where you put it.

Poor memory is the result of lack of attention, or divided attention at the time the particular attention-stimulus knocked. You asked me to buy a ribbon of a certain shade and a certain width when I went to town. I was thinking of my dentist appointment. However, I heard your request, answered it graciously, took the money you offered, still wondering if the dentist would have to draw that tooth. And the chances are that I forgot your ribbon. I was giving you only a passive and divided attention.

Or I have more to do than I can possibly accomplish in the next six hours. You ask me to buy the ribbon. I attend accurately for the moment, think distractedly, "How can I do it all?—but I will"—and crowd the intention into an already overburdened corner of my mind, fail to associate it with the other thoughts already there, and return six hours later without the ribbon. My sense of hurry, of stress, of the more important thing to be done, or a reaction of impatience at the request, forced back the ribbon thought and allowed it to be hidden by others. I was really giving you only partial attention, or an emotion interfered with attention; and I forgot.

Hence we find that a faulty memory may exist in an otherwise normal mind when poor attention, or divided attention due to emotional stress or to an overcrowded mind, which makes it impossible to properly assort its material, interferes.

Again, we forget many things because they are unpleasant to remember. We have no desire, no emotional stimulus to make us remember; or because some of the associations with the forgotten incident are undesirable. We forget many things because if we remembered them we would feel called upon to do some unpleasant duty. You forgot your tennis engagement with B, perhaps, because you were so engrossed in a pleasure at hand, or in your work, that anything which interrupted was, under the circumstances, undesirable. You may have wanted very much to play with him, but some more pressing desire—to care well for your patient, or to continue the present amusement—was stronger. Or you forgot because you did not want to play with him and had no excuse to offer at the time. You wished to forget. Perhaps he does not play a good game, or you do not like him, or at least you like some one else much more, and he happened along; so you forgot B. The unconscious mind saw to it that something else was kept so prominently before your attention that it could not return to the less desired.

Thus a forgetting may be purely the result of an emotional interference which makes it, all in all, more pleasant to forget than to remember. If we would help ourselves or our patients whose memories are faulty, and who make them worse by their continual fretting over their disability, we must train ourselves to be willing to forget all that does not in the least concern our interests or those of the people about us, and does not add anything desirable to our knowledge. Thus we may avoid overcrowding the mind. But when we would remember let us give our whole active attention at the moment of presentation of the new stimulus, and immediately tie it up with something in past experience; let us recognize what it is that we should remember, and call the reinforcement of will, which demands that we remember whether we want to or not. Sincere desire to remember will inspire early and frequent recalling, with various associations, or hooks, until the impression becomes permanent. The average patient's poor memory is made worse by his agitation and attention to it, and his conviction that he cannot remember. The fear of forgetting often wastes mental energy which might otherwise provide keenness of memory. If the nurse ties up some pleasant association with the things she wants the sick man to remember, and disregards his painful effort to recall other things, then—unless the mind is disordered—he will often find normal memory reasserting itself.

We shall consider this question of memory in more detail in a later chapter of practical suggestions for the nurse.

The Place of Emotion

Feeling Cannot Be Separated from Thinking.—Emotion we found the constant accompaniment of every other mental activity. It is first on the stage of consciousness and, in the normal mind, last to withdraw.

When I am working at a problem in doses or solutions, trying to learn my materia medica, or wrestling with the causes of disease in my medical nursing, or thinking how I can eke out my last ten dollars till I get some more, I am pursued with some vague or well-defined feeling of annoyance or satisfaction, of displeasure or pleasure. If all goes well, the latter; if not, the former.

Feeling Cannot Be Separated from Will.—I cannot will without a feeling accompaniment, pleasant or unpleasant. I may be using my will only in carrying out what intellect advises. But we found that intellect's operations are always affective, i. e., have some feeling of pleasure or pain. And the very act of will itself is a pleasant one and much easier if it is making me do what I want to do; it is a vaguely or actively unpleasant one if it is making me act against desire. In the end, however, if I act against desire in pursuance of reason or a sense of duty, the feeling of pleasure in the victory of my better self is asserted. And feeling cannot be separated from will.

Feeling Cannot Be Separated from Action.—I cannot do anything without a feeling of comfort or discomfort, happiness or unhappiness. Try it for yourself when you are feeding a patient, making a bed, giving a bath or massage, preparing a hypodermic. Other things being normal, if you are performing the task perfectly, the feeling of satisfaction, of pleasure, of the very ability to work effectively, with speed and accuracy and nicety, comes with the doing. If you are bungling, there is a pervading sense of dissatisfaction, of unpleasantness. In the automatic or semi-automatic action a great economy of nature has conservatively put feeling at the absolute minimum; but it has not eradicated it. As you walk across the ward, though your predominating thought and feeling may be elsewhere, there is a sense of pleasure or displeasure in the very movement. If your body is fresh and you are of an energetic type and in happy frame of mind, a pervasive feeling of satisfaction is experienced. If tired or discouraged or sore from unaccustomed exercise, every step registers protest.

Thus we find by experiment that there is no thought we have, no single conscious movement or action, nor any expression of the will, but is accompanied with what the psychologist broadly terms pleasure or pain. So emotion, the first expression of mentality, is never absent from any mental or physical act. It permeates all we do, as well as all we think and will, with the partial exception of automatic action, above indicated.

The Beginning of Reason

We found feeling by far the strongest factor in producing action in babyhood and childhood. Our instinctive doing, we learned, is the result of a race impulse. Will acts chiefly at emotion's bidding. But very early the baby's experience operates as a partial check to feeling's exclusive sway. It keeps him from touching the fire, no matter how its brightness attracts. It may be merely the sense memory of hurt when fingers and that bright thing came together; and one such impression will probably prevent him from ever again touching it. Or it may be the brain-cell's retention of the painful feeling of slapped hands when the fingers reaching out to the flame had not yet quite touched. These punishment experiences are only effective in many children after more or less repetition has set up an automatic prohibition from brain to motor nerves; but right here intellect begins to assert itself in the form of sense memory. The baby does not reason about the matter. His nerve-cells simply remember pain, and that particular brightness and glow, and finger touch—or that reaching out to the glow—and slapped hands, as occurring together. In the same way he early connects pleasure with the taste of certain forbidden things. He does not know they are sweet. He only knows "I want." Even here his desire to taste may be checked in action by a vivid memory of what happened when he tasted that other time, and was spanked or put in his little room all alone with only milk and bread to eat for a long time.

Later on the child may think, from cause to effect, thus: "Sweet, good, want, taste, spank, hurt (or no dinner, all by self, lonely), spank hurt more than sweets good. Not taste." But long before he can work this out, consciously, two distinct memories, one of pleasure and one of pain, are aroused by the sight of the sweet. And what he will do with it depends upon which mem ory is stronger. In other words, his action is governed altogether by his feeling, though memory, which is an intellectual factor, supplies the material for feeling.

Development of Reason and Will

Later still, when the child is older, we may have somewhat the following mechanism: "Sweets, good, want, taste; spank, hurt; don't care, spank not hurt much, maybe never found put, sweets very good."

Now the child is reasoning and choosing between two courses of action, don't and do. His decision will depend upon whether immediate satisfaction of desire is stronger than the deferred satisfaction of being good, and the fear of punishment. He probably prefers to take a chance, and even if the worst comes, weighs it with the other worst, not having the sweet—and takes the "bird in the hand." He has reasoned, and has chosen between two emotions the one which his judgment says is the more desirable; and his will carries out the decision of his reasoning. His chief end in life is still to get the most immediate pleasure. Still later in child-life, much later, perhaps, his decision about the jam is based on neither love of it nor fear of punishment, but—despite his still sweet tooth—on a reasoned conclusion that if he eats jam now he may be sick, or he may spoil his appetite for dinner; or on a consideration that sweets between meals are not best on dietetic principles; and will very readily backs up the result of his reasoning. Though his determination is largely based upon feeling, reason has chosen between feelings, between immediate desire to have, and desire to avoid future discomfort. Reason is triumphant over present desire.

Judgment

The conclusion or decision that reason has reached we call a judgment. The youth who decides against the sweet between meals, we say, has good judgment. And we base our commendation on the proved fact that sweets are real fuel, giving abundantly of heat and energy, and are not to be eaten as mere pastime when the body is already fully supplied with high calorie food not yet burned up; that if sweets are eaten at irregular intervals and at the call of appetite, and not earned by an adequate output of physical work, the digestive apparatus may become clogged, and an overacid condition of the entire intestinal tract threaten. We call judgment good, then, when it is the result of reasoning with correct or logical premises which correspond with the facts of life. We call it bad when it is the conclusion of incorrect or partial or illogic premises.

A premise "is a proposition laid down, proved, supposed, or assumed, that serves as a ground for argument or for a conclusion;

a judgment leading to another judgment as a conclusion" (Standard Dictionary).

Let us illustrate good and bad judgment by following out two lines of reasoning, each quite accurate as such.

I want sweets. Sweets are good for people. They give heat and energy, and I need that, for I am chilly and tired. People say "Don't eat sweets between meals." But why? They contain just what I need and the sooner I get them the better.

So I have sweets when I want them. The judgment to take the sweets as desire indicates is entirely logical if we accept all the premises as correct. And they are, so far as they go; but they are partial; and so cannot altogether correspond with the facts of life. Sweets are good for people who expend much physical energy. They prove injurious in more than limited amounts to the bed-ridden, the inactive, or the sluggish. Hence this premise is partial and so far incorrect. Sweets do give heat and energy, true. I am chilly and tired, also true. But why? Because I am already toxic from the sweets and meats I have had throughout my sedentary years. The question is, Do I need any more energy-producing food when I am not burning up what I have? So again the premise is partial. I do need heat and energy, but I already have the material for it, and my mode of life has disorganized my system's capacity to utilize these foods normally. So now sweets have become a detriment to my well-being. The judgment which determines me to the habit of eating sweets between meals is the result of logic, but of logic spent on tying up premises which do not fit the facts of the case.

One of the most prevalent defects of judgment is illustrated in this common disability to select premises which fit the facts. Ignorance, emotional reasoning, and a defective critical sense probably explain most poor judgments.

The other judgment illustrates the logic of correct, provable premises.

"No, I shall wait until dinner-time. I have no need of so rich a food, for I had an adequate meal at the usual time and have not worked hard enough to justify adding this burden to my digestive apparatus; besides only hard workers with their muscles can afford to eat many sweets. They cause an overacid condition when taken in excess; and any except at mealtimes would be excess for me, with my moderate physical exercise."

This judgment we call good. Its premises correspond to scientific facts.

But much reasoning must always be done with probable premises, ones which seem to correspond to the facts, but which have yet to be proved. And our judgment from such suppositions cannot be final until we see if it works.

Some few centuries ago supposedly wise men called Christopher Columbus a fool. Of course the world was flat. If it were round man would fall off. It was all spread out and the oceans were its limits. If it should be round, like a ball, as that mad man claimed, then the waters must reach from Europe 'round the sphere and touch Asia; or there might be land out there beyond the ocean's curve. But it wasn't round, and the idea of finding a new way to Asia by sailing in the opposite direction was a fool's delusion.

Their logic was perfect. If the earth was flat, and Asia lay east of Europe, it was madness to sail west to reach it. But they argued from a wrong premise, so their judgment was imperfect—for they did not yet know the facts.

The result of all reasoning is judgment. And judgment is good as the materials of the reasoning process correspond to facts, or are in line with the most probable of the yet unknown. It is poor as the reasoning material fails to meet the facts, or is out of harmony with the most probable of the yet unproved.

It is of no avail, then, to attempt to improve our final judgments as such. We must examine the materials we reason with, then learn to group and compare them logically. And in the very separating of true premises from false, we use and train the judgment we would improve. And this the normal mind can do.

Reaction Proportioned to Stimuli

In the normal mind the emotional or feeling accompaniment of thought and action is proportionate and adequate to the circumstances, i. e., there is a certain feeling, of a certain strength, natural to every thought and act; and when only that strength, not more or less, accompanies the thought or the act, we say, "That man is emotionally stable. His mind is normally balanced."

Joy naturally follows some stimuli; sorrow others. Disappointment or loss, shock, failure, death of loved ones, illness in ourselves or others, do not normally bring joy. A keen sense of suffering, temporarily, perhaps, of numbness; the inability to grasp the calamity; or flowing tears, an aching heart, or the stress of willed endurance, are natural, and normal reactions to such stimuli.

A developed will may refuse indulgence in the outward expression of the normal feeling of shock, grief, and loss; and this may be normal. But normal volition does not force us to laugh and dance and be wildly merry in the face of grief and loss and pain. It only suggests the adequate, reasonable acceptance of the facts that cannot be changed — the acceptance of love, faith, and hope that sees in present suffering a means of consecration to service; it does not convert the emotion of sorrow and loss into a pleasurable one. Normal reason does not suggest that will force the reactions to loss and suffering that belong by nature to attainment and success.

Nor does reason suggest the long face, the bitter tears, a storm of anger, in response to comedy and farce, in the face of a good joke, or to meet success; and normal will puts reason's counsel into effect.

Normal Emotional Reactions

Some emotions, that seem exaggerated at first thought, may be normal under the circumstances. For no one can know the whole background for emotional response in the life of another. After being long shut up in a darkened room, with bandaged eyes and aching head and sick body, the first visit to the bit of woods back of the house—when all the pains have gone—may bring almost delirious joy. The green of the foliage, the blue of the sky, the arousing tang of the air, the birds, the sense of freedom—all go to the head like new wine. The abandon of joy is a normal response under the circumstances, now. It would hardly be normal to one whose habit it is to visit this same bit of woods every day, to one who loved it, but for whom it had lost the force of newness.

To the child, who has never in all his little life had a wish not gratified, the denial of a desired stick of candy is as great a calamity as is the loss of a fortune to the grown man. And the child reacts to feeling equally intense. These are normal reactions to stimuli—normal, under the circumstances.

The Normal Mind

The normal mind reasons clearly with the best data at hand to results that will stand the test of conformity to reality; the normal mind uses reason and feeling, guided by reasonable attitude; in the normal mind reason advises action and will brings it about; in the normal mind feeling proportionate to the circumstances accompanies every thought and every action. And in the well-balanced man or woman every function of the mind leads to action as its final end.

But man only approximates the normal. The perfectly balanced man or woman is so rare as to be a marked person. The average intelligent individual only in general approximates this standard. He goes beyond it in spurts of untrammeled genius, to wrench lightning from the heavens, and to send his trains through the air; or he allows his feelings to dictate to his reason, and much of the time so exaggerates or depreciates the simple facts of life that the results of his reasoning no longer conform sufficiently to reality as to be thoroughly dependable.

CHAPTER VII
PSYCHOLOGY AND HEALTH

In the use of its functions the mind manifests certain powers and certain modes of expression which can act as powerful allies or as damaging enemies of health. We speak of man as adaptable, but also as a being of habits. We speak of him as "feeling" when we wish to express the fact that his emotions influence his body. We expect of the average man a certain amount of suggestibility. We say that he is tremendously affected by his environment, which simply means that his attention, naturally centered chiefly on the things at hand, largely determines what he is. But we recognize that a man of trained mind can choose and will to substitute for his present surroundings thoughts upon more constructive things from past experience, or from future possibilities, or from within the mind's own storehouse. His ability to largely modify his life by his will, we recognize as man's greatest power. Adaptability, emotional response, suggestibility, attention, thought-substitution, habit-formation, and will can minister vitally to health, or can prove damaging avenues of disease.

Necessity of Adaptability

Adaptability is as essential to life of mind as to life of body; and health of mind as well as health of body is determined by the individual ability to adjust himself to environment.

There are dreamers who have lived in their ideal world so long that they cannot meet the stern realities of life when they come. The shock is too great for the mind that has accepted only the fantastic, the real as the dreamer would have it; and he lets go altogether his hold on the actual, accepting the would-be world as present fact. And we call him insane. Other visionaries wakened rudely to life as it is, accept it as unchangeable fate, lose all their true ideals and become cynical, or victims of utter depression for whom life holds nothing that matters. Still others go on through the years self-satisfied and serene because they simply refuse to believe unpleasant truths; they "pretend" that their wishes are realities, and acknowledge as facts only the pleasant things of existence. The first two groups have failed to adapt self to life as it is, and the mind is lost or so damaged as to no longer serve its body properly. The "pretenders" have adjusted themselves, and so long as they can remain happily self-deceived all goes well for them, though they complicate living for others. However, they have made an adaptation, a defective one, it is true, but one through which the mind may survive. Some of this class, however, finally build up a more and more elaborate system of self-deception until they, too, are insane.

The practically adaptable man can dream dreams, but always recognizes them as dreams, and can stop at will; can vision a beautiful ideal, but comprehends that it is not yet reality, though it may some time become so if he learns and fulfils the laws leading to its realization. The adaptable man or woman recognizes the real as fact, desirable or otherwise, the fantastic as unreal and only to be indulged in as a pastime, and the ideal as the possible, a thing for which to work and sacrifice. So perfect adaptability would mean perfect mental poise.

It is for the nurse to realize that the greater number of her patients do not belong to any of these classes absolutely, but that some of them have tendencies leading in these various directions. And it is her privilege to recognize the trend of her sick patient's mental workings, and to so deftly and unobtrusively encourage the recognition of facts as things which are to be used—not as stumbling-blocks—that her mental nursing, as her physical, shall be directed toward health. She can help her patient to accept illness and suffering as realities to be faced, and treatment as a means, whether pleasant or not, of making it possible for health to replace them. The understanding nurse can actively help her charge one step at a time toward adaptation to the new environment, remembering that many of the sick, particularly the depressed, cannot be encouraged or incited to effort by having future health held out to them. They are capable only of living in the present and doubting all the future.

There Can Be No Neurosis Without a Psychosis.—If the brain is the organ of the mind, then what affects the brain must perforce be at least registered by mind. So every physical shock, accident, toxic condition, infection—even the ordinary cold—rouses the mind at least to awareness, usually to discomfort. For the nerve-cells and fibers—those inseparable parts of the body mechanism—speedily report the fact that they are being tampered with. In the toxicity of the infections these very delicate tissues are nourished by toxic fluids; in accidents they carry all the messages from the injured part. Then the brain—that center of all man's reactions and the organ of all his consciousness—receives the report of the disturbance and translates it into terms of more or less disability. The neurosis has become a psychosis. The physical condition has become a mental discomfort. Normally this ensuing mind state should be in accordance with the extent of the injury to the nerve-cells and fibers. But under long-continued discipline, or influenced by emotion, the conscious mind may not recognize the neurosis; whereas, in the hypersuggestible, consciousness will translate it into entirely disproportionate suffering.

A great problem of nervous education is what the mind will do with discomfort or pain. Will it put all its attention there and respond with nervousness, irritability, demand for sympathy; or will it relegate all the minor pains to their own little places, accepted as facts but to be disregarded except in so far as actual treatment is needed? Will it turn to attend to the host of other more desirable objects? Or in case of acute suffering, will it take it as a challenge to endurance? Will it use it as a means to strengthen volition, as a stepping-stone to self-mastery?

Realizing the force of the law—no neurosis without a psychosis—

the nurse will try to eliminate unnecessary irritations to physical comfort, while she helps the patient to adjust himself to the ones which are inevitable. It is the doctor's problem rather than hers, except as she carefully fulfils orders, to eliminate the toxic causes of psychosis. It is hers to help the patient to meet adequately the effects of the infections or toxins, and to prevent as far as possible the surrender to uncontrolled nervousness. Her object is to have him face the psychosis as one of the simple facts of science, then turn the sick mind's attention to more important things; she would encourage will to force endurance; she would stimulate the feeling life to the forward look of confidence and faith, or to acceptance of life's suffering as a challenge. The nurse knows that pains beyond the power of endurance the doctor will lighten. And the patient's reaction to discomfort and suffering, the understanding nurse, without any preaching, can very largely influence.

The Power of Suggestion

One almost universal condition found in illness is hypersuggestability. Here is the nurse's despair and her hope. Suggestion may come from without or from within. When from within, we call it autosuggestion.

Many of the sick are temporarily resting their reasoning faculties and their judgment. The sick body is causing a feeling of "jangling nerves," and the mind, too, is strongly tempted to be sick. So every harsh sound, every jolt, almost every sentence spoken in their hearing suggests immediate nervous reactions. The mind does not wait to weigh them. The nervous system reacts to them the second the impression is registered. The whole self is oversensitive, and the very inflection of a voice has enormous significance. Let the nurse remember that her way of giving a treatment, her expression, or her very presence becomes a potent stimulus on the second, one to which the patient's mind responds like a flash-light when the button is pressed.

The nurse must comprehend the principle of the nervous effect on the patient of all that is done and said, and realize her tremendous privilege in making those stimuli wholesome. The nurse who has a sympathetic insight, with unswerving loyalty to orders, can carry them out with the average patient, unpleasant though they may be to him, in such a way that his wholesome emotional response will be called forth, a response of co-operation, or of faith or of good breeding, or of "downing" the impulse to indulgence; or a response directed toward holding the nurse's interest and attention, and so keeping her in the room; such a response as will gain some privilege, etc.

But there are some patients in whose cases ordinary persuasion, suggestion or requests fail. They are too nervously or mentally sick to be moved by logic, or to respond with customary grace to a request which their reason is not awake to answer. All usual suggestions may fail of effect. And for these few, in order that health may be at all assured, even the discipline of force may be necessary. But the nurse must use this only as a last resort, of course, and in accordance with the doctor's orders, and then solely as treatment leading toward the ways of health. Before turning to this final method she should clearly, firmly, and kindly explain the principle of the discipline if the patient's mind is at all capable of grasping it. In any case, force should be used only as the surgeon uses his knife. It hurts, but only to help and to save; and it is not called upon when other methods can secure the needed results. But force, thus limited in its application, may prove the only suggestion which will bring about the action necessary to health on the part of the patient. Force unwisely and unkindly used proves a damaging suggestion, causing reactions of fear or anger; or it may lead to delusions of persecution and to strengthened resistance.

Many suggestions come to the patient from within. Discomfort in the right side may suggest appendicitis. A slight indigestion, often purely nervous, may be interpreted as inability to care for certain diet, etc. The wise nurse will displace as many of these as she can by casual suggestions on her own part. She will demand of herself that her very presence be quieting, calming, happy; that her conversation with her patient shall vibrate with a certain something that gives him courage and strengthens the desire and the will to health; that her care of him shall prove confidence-breeding. The patient's attitude, when he is at all suggestible, is largely in the nurse's hands, and she can make his illness a calamity by dishonest, fear-breeding, or suspicion-forming suggestion. After all, the whole question here is one of the normality of the nurse's own outlook on life and people. The happier, truer, and more wholesome it is, the more really can she help her patient to both bodily and mental health. Of one thing let the overzealous nurse beware. Do not irritate your patient by a patent, blatant, hollow cheerfulness that any one of any sense knows is assumed for his benefit. Personally I know of no more aggravating stimulus.

What We Attend To Determines What We Are.—This is one of the first laws of education. If the child's attention from birth could be controlled, his future would be absolutely assured. But attention is a thing of free will and cannot be forced by others. It can be won through interest or self-directed by will. The child's attention is entirely determined by interest, interest in the morbid and painful as truly as in the bright and happy. Punishment interests him tremendously because it affects him, it interferes with his plan of life, it holds his entire immediate attention to his injured self. But something more impelling quickly makes him forget his hurt feelings and he is happy again. The average sick person is emotionally very much like the child. His will at the time, as we noted before, is tempted to take a rest, and his interest is ready to follow bodily feeling unless something more impelling is offered. The nurse who can direct attention to other people, to analyzing the sounds of the street, to understanding something of the new life of a hospital or sick room, to planning a house, or choosing its furniture or equipping a library, or supplying a store; to intelligent references to books or current events; or to redecorating the room— all in his mind; to an appetizing tray, a dainty flower, a bit of sunshine, a picture, etc., is fixing the patient's attention on something constructive, helping him to get well by forgetting to think of himself.

Thus the nurse, knowing the laws of attention, can keep herself alert to divert and direct her patient's thought to wholesome interests. Knowing the possibility of thought substitution, she can open up new channels of thinking. Knowing the power of the will to assist in health bringing and health keeping, she can sometimes stimulate long-dormant determination. Let her beware, however, of making the convalescent too dependent upon help from without, but prick his pride to gradually increasing doing for himself. Arouse his reasonable ambition, but let him realize that life must be taken up again a step at a time; and that he can do it. If limitations must be accepted, try to inspire the feeling of pride in accomplishing the utmost possible within a limitation, and an acceptance of the inevitable without bitterness.

Attending to the unhappy, the painful, the boring without looking beyond makes life unhappy, painful, and a bore. Not that the nurse should ignore these realities, but she can accept them wholesouledly herself as not the final things, as merely the rocks that can be used to stand upon and get a view of the something better for everybody. When they are thus used by the wholesome mind, facts, the very barest and meanest of them, can be made useful as stepping-stones to the happier facts beyond them.

If the nurse can direct or tactfully lead the patient's attention away from himself and his illness, she has found a big reinforcement to his treatment. This question is so vital in the care of patients that it will be discussed at greater length later on.

One Thought Can Be Replaced by Another

If we control attention we control thought, and with the suggestible patient this principle depends upon the one just now considered. Hope and courage-breeding thoughts can replace despairing and fearful ones, but it will be only when attention is directed through interest or by will to new material. There is no blank in waking consciousness. The last thought or feeling or perception, through association of ideas, brings up a related one, and so on indefinitely. We may start with a pebble on the road and go on logically, smoothly, until in five minutes we are thinking of the coronation of King George, with no sense of anything at all unusual in the succession. It may be a very roundabout process, from "pebble" through "rough way," "ways that hurt," "dangerous ways," "brigands," "uncertainties of life." "Uncertain lies the head that wears a crown," "King George and his crown," "coronation." But this constant stream of thought can be broken into at any point by a spoken word, a passing vehicle, which diverts the mind's trend. So the nurse can take advantage of the mind's very suggestibility, and substitute for the unhappy and sickness breeding by turning attention to anything else of a happier color, and may divert the entire stream of thought in that direction. She who knows these simple laws of the mind, and who at all knows people, is a therapeutic agent of unlimited value.

Habit is a Conserver of Effort

It is always easier to follow a beaten path than to break one's way through untrodden forests. It is easier to walk after we "learn how," and learning how is simply doing it over and over until the legs and feet have acquired habits of motion and accommodation to distances and to what is underfoot. It is easy to do anything after we have done it again and again, so that it has become second-nature, and "second-nature" is habit. The wise man early forms certain habits of personal care, of eating, sleeping, exercising; of study, of meeting the usual occurrences of life. The first day he spent at anything new was a hard one. Nothing was done naturally. Active attention had to be keenly held to each detail. He had to learn where things belonged, how to do this and that for the first time, how to work with his associates.

Do you remember the first hospital bed you ever made, the first bed-bath you gave, the first massage? You had to be taught bit by bit, detail by detail. You did not look upon the finished whole, but gave almost painful attention to each step that led to the made bed, the completed bath, or the given massage. Your fingers were probably all thumbs unless you had experience in such things before you came to the hospital. Your mind was tired from the strain of trying to remember each suggestion of your instructor. The second time, or certainly the third or fourth time, it went better. After a week of daily experience you gave the bath or massage or made the bed with much less effort. A month later the work was practically automatic and accomplished in a fraction of the time you spent on it that first day. Now you can do it quickly and well with little conscious thought; and at the same time carry on a brisk conversation with your patient or think out your work for the day. Your mind is free for other thoughts while you perform the task easily and perfectly. Your method of doing the work has finally become a habit which saves the effort of conscious attention. The details of your routine work are directed by the subconscious. The habit will be energy and time saving in proportion to the accuracy of your first conscious efforts spent on the new undertaking. Thus, useful habit is the result of active effort.

We can acquire habits of thinking and habits of feeling as well as habits of doing.

But the other habits, the bad ones, are not acquired with effort. We fall into them. Hazy thinking is easier than clear thinking. Suppose you are by nature rather oversanguine or overdespondent, and you make no genuine attempt to evolve that nature into poise. Directing will to do what desire opposes is too difficult, and you go the way of least resistance. So easily are the bad habits formed; but only with tremendous effort of will and persistence in refusing their insistent demands can they be broken or replaced by helpful ones.

But habits can be learned; and bad habits can be broken when an overpowering emotion is aroused against them, possesses the mind, and controls the will; or when reason weighs them in the balance and judgment finds them wanting, and volition directs the mind to displace them by others.

The nurse meets in her patients numberless habits which retard recovery of body and make for an unwholesome mental attitude. Some patients have the complaint habit, some the irritation habit, some the self-protection habit, some the habit of impatience, some of reckless expression of despair, some of loss of control, some of incessant self-attention. The nurse who can arouse an incentive to habits of cheer expression when the least cause of cheer appears, who can by reason, or if that is not possible, by suggestion; by holding out incentives, or by making some privilege depend upon control—this nurse can help her patient to displace habits of an illness-accepting mind by habits of a health-accepting one. Above all, let her beware of opening the way to habits of invalidism. Some people acquire the "hospital habit" because it is easier to give way to ill-feeling, however slight, and to be cared for with comfort, than to encourage themselves to build up endurance by giving little attention to minor ailments.

The Saving Power of Will

It is not uncommon to hear a doctor say, "Nothing but his will pulled him through that time." It does not mean quite what it says, for the patient's will would have been helpless to cure him without the medicine and the treatment. But it does mean that in some cases when life is hovering on the brink, even the most skilful treatment cannot hold it back if the will to live is gone. The chances may be half and half. Lack of desire to live may drop the balance on the death side. Determination and hope and confidence may overweigh the life side. For the influence of will in refusing to surrender to depression may throw the needed hair's weight in favor of more normal circulation. Depression and emotion may so effect the sympathetic nervous system as to cause a lowered circulatory activity. Determination, based on volition, may stimulate a response from the sympathetic system which will increase heart activity. And certainly, when it is not a matter of life and death, but a prolonged recovery, will is a saving grace. The patient who sets all his sick energies to the task of winning health reaches his goal quicker than the hopeless and depressed. Perhaps his will merely brings utter relaxation for the time, forces acceptance of present helplessness only for the sake of giving the body a better chance to recuperate; but the very fact that it is acting to hopefully carry out orders lightens by half the nurse's task of getting him well; and she can encourage this will to co-operate with the doctor's efforts by suggestion, by her directness and honesty, by the quiet assurance that at least a reasonable degree of health is won by effort.

We have touched upon only a few of the laws of the mind. The nurse can help develop saving mental habits and wholesome attitudes while she helps to strengthen sick bodies; she can make a cure a little more certainly lasting who will remember that:

1. Adaptability is essential to life and health.
2. There is no neurosis without a psychosis.
3. Suggestion may be a powerful factor for health.
4. What we attend to determines what we are.
5. Thought substitution is possible.
6. Habit is a conserver of effort.
7. Will is a saving power.

CHAPTER VIII
VARIATIONS FROM NORMAL MENTAL PROCESSES

Disorders and Perversions

Life would be a very simple proposition if the mental machinery always worked right. But this is peculiarly subject to damage both from without and from within. From without it may be damaged by the toxins of food, as in the acute toxic psychoses; by the poison of drink, as in the alcohol-produced psychoses, such as acute alcoholic hallucinosis; by lack of muscular exercise, resulting in a deficient supply of oxygen to burn up the accumulated toxins from energy-producing foods; by the infections, which may result in the infection-exhaustion psychoses; by wrong methods of education, and by surroundings which demand too severe a mental strain in the struggle toward adjustment. These damages from without we class roughly as environmental.

From within the mental workings may be injured by emotional dominance; by bad habits of thinking and feeling and doing—often the result of wrong methods of education; by defective heredity; by undeveloped will; by the insanities. These danger sources from within we might classify as self-produced and hereditary.

There may be disorders of any or every function of the intellect, disorders of feeling, and perversions of will. Some of the most commonly met we list below.

- Disorders of Sensation

Hyperesthesia (exaggeration of sensation)
    as found in neurasthenia, or in mania.
Anesthesia (absence of sensation)
    as in the numbness of hysteria; in sensory paralysis.

Retardation
> as in dementia and melancholia.

"Clouding" or dulness
> as in simple depression.

Perversion
> as in dementia and melancholia. Sweet may taste sour; fresh food may smell decayed.

- Disorders of Perception (being dependent on sensation is always disturbed with it).

Hyperesthesia (exaggeration)
> as in neurasthenia or mania.

Anesthesia
> as in hysteria or paralysis.

Retardation
> as in dementia and melancholia.

"Clouding" or dulness
> as in simple depression.

Illusion
> found in normal mind — easily corrected;
> found in many insanities.

Hallucinations
> frequently met in the infection-exhaustion psychoses, in dementia, in paranoia, in acute hallucinosis of alcoholism.

- Disorders of Ideation

Hypochondriasis
> found in many of the hypersuggestible, frequent in the mild depressions and in all victims of self-attention.

Retardation
> found in most depressions.

Deficiency
> as in idiocy — the inability to form new concepts.

Acceleration
> as in hypo-mania.

Poverty
> as in the abnormally self-centered;
> as in melancholia.

Rambling ideas
> as in chronic insanity.

Flight of ideas
> as in manias, hysterias, and acute deliriums.

Fixed ideas
> as in paranoia.

Perversions (concepts change their meaning altogether)
> as in dementia.

Ideogenous pains
> as in hysteria.

Compulsive ideas
> common in borderland states;
> in psychasthenia, or hysteria.

Disorientation
> (wrong idea of { thing, place, or person);
> found in confused conditions;
> in delirium from infections;
> in insanities.

Confusion
> as in the infection-exhaustion psychoses;
> in insanities.

- Disorders of Memory

Absent-mindedness.

Amnesia (morbid forgetfulness).

Aphasia { temporary, prolonged, permanent (see later explanation).

Perversion
    as fabrications, due to memory-confusion or inaccuracy; also due to excessive ideation and defective judgment.

- Disorders of Reason

Delusions
Systematized Transient Fixed
Somatic
    as in hypochondriasis.
Persecutory
    as in paranoia.
Unworthiness
    as in simple depression or melancholia
Grandeur
    as in mania or paranoia.
Nihilistic
    often found in melancholia.
Reference
    as in paranoia.
Altered personality
    as in hysteria.
Perverted personality
    (patient may believe he is a dog);
    as in dementia.
Emotional thinking.
Shut-in personality

as seen in the deficient social capacity of potential dementia præcox.

- Disorders of Judgement

Defective judgment
>in all insanities;
>in hysteria.
>*Ex.*: Patient who accepts mental suggestion of disability as reality.

Perverted judgment
>in severe dementias—as influenced by unreasonable fear, hatred, etc.;
>in all acute insanities—as manifested in inability of patient to rid himself of his delusions.

Absence of judgment
>in all acute insanities;
>in later dementias.

Limitations
>in many so-called normal and in all the abnormal.

- Disorders of Emotion

Suggestibility
>in hysteria.

Excitement
>in mania.

Depression
>in melancholia.

Phobias
>as found in psychasthenia.

Deficiency
>as in the apathy of depression.

Perversion
>in mania, in depression, in catatonia.

Deterioration
> in dementia.

Sense of unreality
> found in all borderland cases.

- Disorders of Will

Wilfulness
> in many "normal." Very common in hypomania.

Willessness (aboulia or paralysis of will)
> often found in psychasthenia; and in depressive states.

Morbid inhibition
> as in depressive states.

Indecision
> as in psychasthenia;
> as in simple depression.

Obsessions
> found pre-eminently in psychasthenia.

Tics
> in many borderland cases;
> in the hypersensitive as often the only expression of any neuropathic tendency.

Distractibility
> as in hypomania and frequently in hysteria.

Negativism
> as in catatonia.

Mutism
> as in catatonia.

Compulsive acts
> as in psychasthenia, hysteria, etc.

Psychomotor overactivity (volition unable to check)
> as in mania.

Psychomotor retardation (volition unable to energize)

as in depression.

From this limited survey of the mind's disorders we realize that every departure from the normal mental attitude tends to associate itself with one of the following five *states of mental disability*.

- Depression,
- Exaltation,
- Perversion,
- Enfeeblement,
- Deficiency.

CHAPTER IX
VARIATIONS FROM NORMAL MENTAL PROCESSES
(Continued)

Hyperesthesia is abnormal sensitiveness to stimulation.

Anesthesia is loss, either temporary or permanent, of any of the senses.

Perversion is morbid alteration of function which may occur in emotional, intellectual, or volitional fields.

Example: The odor of a rose causing an acute sense of physical pain.

An illusion is a false interpretation of a perception.

The normal mind is quite subject to illusions, either due to a faulty sense organ, or to a preconceived state of mind which so strongly expects or presages something else than reality as to misinterpret what the senses bring.

Examples:　　The crooked stick as a snake.
　　　　　　　A ghost created from shadow.
　　　　　　　An ordinary ringing in the ears as sleigh-bells.
　　　　　　　Milk tasting like blood.
　　　　　　　An hallucination is a perception without an object.

The hallucinated individual projects, as it were, the things of his mind's creation into the outer world, and accepts them as reality. He sees snakes where there is nothing to suggest them; sees a ghost where there is no shadow; believes that the taste of blood is constantly in his mouth.

There are possible hallucinations of every sense. Nonexistent objects are seen, touched, tasted, heard, or smelled.

Hypochondriasis is a state characterized by persistent ideas of nonexistent physical disabilities.

The hypochondriac has every known symptom of indigestion, or of heart disease, or is threatened with tuberculosis—all in his mind; and whatever the disorder he seizes upon, his attention hovers there, while the ideas of that particular disability persist and strengthen.

A flight of ideas is an abnormal rapidity of the stream of thought.

Every perception so immediately is linked with some association of experience that expression is swift and often incoherent. One word will follow another with amazing rapidity, words suggested by sound association, usually, rather than by that of meaning.

Example: "Made a rhyme, had a dime, did a crime, got the time, bring some lime." This association by rhyme is quite common. But the associations of meaning are not uncommon.

Example: "Made a rhyme. Mary was a poet. Mary had a little lamb. Where's Mary?—Mary!—No Jim—Jim, all my children—calling, calling, calling," etc.

A fixed idea is one which morbidly stays in the mind and cannot be changed by reason.

Example: In hypochondriasis, as given above.

Ideogenous pains are either pains born of an erroneous idea, or mental reproductions of pains now having no physical cause.

A suggestible person, learning that his grandfather died of an organic heart, conceives the idea that he has inherited the trouble, and begins to suffer cardiac pains; and as long as the idea persists the pain is felt.

Compulsive ideas are ideas which intrude, recur, and persist despite reason and will.

Example: The compulsive idea of contamination may lead its victim to wash and rewash his hands at every contact with matter, until finally, though they are raw and sore, he is incapable of resisting the act.

Disorientation is a state of mental confusion as to time, place, or identity.

Amnesia is pathologic forgetfulness.

Example: As sometimes found in the infection-exhaustion psychoses, when the entire past of the patient may be wiped out for the time. Cases of permanent amnesia are known.

Aphasia is a defect in the interpretation or production of language.

There may be motor aphasia, auditory aphasia, vocal aphasia, sight aphasia; and with disability to produce words, they may yet be recognized when seen; or when they can be spoken they may not be recognized when heard; or with inability to speak them, they are accurately sensed by hearing; or though understood when heard, they are incomprehensible when read.

A delusion is a false belief which cannot be corrected by reason.

A somatic delusion is one centering upon alterations in the organs or their functions.

Example: Absence of a stomach, inability to swallow.

A nihilistic delusion is one which denies existence in whole or part.

Example: Mother denies the existence of her child.

A delusion of reference is one in which the deluded individual believes himself an object of written, spoken, or implied comment.

Example: The actors on the stage are directing their remarks directly against the victim in the box.

A shut-in personality is one that habitually responds inadequately to normal social appeal.

Sense of unreality is one of the commonest psychic alterations through which customary sensation states are displaced by unnatural and usually distressing ones.

Examples: The breakfast table appears undefinably altered.
Laughter is accompanied by strange, rather than by normal, sensations.
Morbid inhibition is an abnormal, negative activity of the will.

Sometimes a patient will try pitifully to express some thought or feeling; the desire to explain is there, but will is blocked in action. Or the patient attempts to dress, makes repeated new beginnings, but cannot succeed. We say, "He is inhibited."

An obsession is an idea which morbidly dominates the mind, constantly suggesting irrational action.

Obsessed patients may consistently step in such a way as to avoid the juncture of the flagstones on the pavement; may insist on removing their shoes in church; may hail each person met on the street and tap him on the arm; may refuse to ever leave the house without an open umbrella; or may try to attack every man they see, not because they want to hurt or kill, but because they are obsessed to the performance of the action.

A tic is a useless, habitual spasm of a muscle imitating a once purposeful action.

Motor tics, such as habitual jerking of the arms, shrugging the shoulder, contorting the face, shaking or nodding the head, snapping the fingers, etc., are very common among nervous children, and even in many otherwise normal grown-ups.

Distractibility is an abnormal variation of attention.

The common inability of the hypomanic patient to hold his attention to any subject when another is open, is very like the distractibility of the child who turns to every new interest as it is presented.

Negativism is a state of persistent compulsion to contrary response to suggestion.

It is with these patients as though not only initiative were lost but also the power to follow another's lead. But their independence asserts itself in opposing every suggestion and in acting so far as possible contrary to it.

Mutism, as used in psychiatry, is an abnormal inhibition to speech.

Patients sometimes speak no word in many months. To all appearance they are true mutes. Then suddenly something may remove the mental blockade and they talk.

Compulsive acts are acts contrary to reason, which the will cannot prevent.

A seemingly quite normal patient will sometimes grab a vase from a stand in passing, and dash it to the floor. Something "urged" him to do it, and he could not resist. Others will tear their clothes to shreds, not in anger, but because they "could not help it."

Psychomotor overactivity is abnormal activity of both mind and body, contrary to reason and uncontrolled by will.

Psychomotor retardation is an underactivity of both mind and body in which consciousness is dulled and the body sluggish.

A neurosis is a disorder of the nerves, which may be functional or organic.

Nervousness is properly termed a psychoneurosis—for we have learned that there can be no neurosis without an accompanying psychosis.

Psychosis is the technical synonym for insanity.

Borderland disorders constitute a group in which mental perversions do not yet so dominate reactions as to make them irrational.

Twilight is neither night nor day; the feelings of the hysteric are not insane, but the actions may be.

Insanity is a prolonged departure from the individual's normal standard of thinking, feeling, and acting.

Mania is insane excitement.

Melancholia is the inability of the mind to react to any stimulus with other than gloom and depression.

Melancholia may be of the intellectual type or of the emotional type. The patient who tells you constantly that he has murdered all his children, that he is a criminal beyond the power of God to redeem, who seems chained to his delusions, yet shows no adequate feeling reaction, no genuine sorrow, we call a case of the intellectual type of melancholia. Another patient misinterprets every normal reason for happiness until it becomes a cause of settled foreboding. The mother, whose son fought safely through the war and is now returning to her, feels that his coming forecasts calamity for him. He had better have died in France. She is of the emotional type of melancholia.

Hysteria is a nervous disorder based upon suggestibility, and capable of imitating most known diseases.

Insane impulses are morbid demands for reckless action beyond the control of the will.

Example: The impulse to kill, quite regardless of who may be the victim.

Psychopathic personality is a term much used today to designate an hereditary tendency on the part of the individual to mental disorder.

The neuropath is the individual with an inborn tendency to the neurosis.

Neurotic is a term broadly employed for the nervous in whom emotions predominate over reason.

Neurasthenia is a nervous disorder characterized by undue fatiguability.

Psychasthenia is a nervous disorder characterized by a sense of unreality, weakness of will, self-accusation, and usually by phobias and obsessions, all subject to temporary correction by reason or influence from without.

Hypochondriasis is a disorder characterized by morbid attention to bodily sensations, and insistent ideas of bodily disorder.

Phobia is a morbid fear or dread.

FACTORS CAUSING VARIATIONS FROM NORMAL MENTAL PROCESSES

Heredity

When we consider the accumulated possibilities for disorder which the family tree of almost any one of us can show, the wonder is not that there are so many nervous or insane, but rather that any come within hailing distance of the normal. For multitudes are born of parents whose bodies were food poisoned or alcohol or drug poisoned, and whose nervous systems were tense and irritable, oversensitive, and suffering from the effect of these same toxins on the brain. Others are of manic-depressive parentage; some are possibly even of paranoic or dementia præcox lineage; while many of our finest and best had psychopathic or neuropathic heredity. Syphilis, itself, and the underpower bodies of tuberculosis are heritages of many.

When we realize, too, that we are born with certain inherent tendencies of temperament, which are too often of the melancholic or overcholeric type, our wonder grows that we are not doomed to defeat at birth. Were it not for the possibilities in the germ-plasm of choosing the much of good also in our heredity, often enough to overbalance the bad, and for the proved power of environment and training to modify or even altogether overcome the harmful parts of our birthright, there would be little hope for many.

Environment

While environment may prove the saving grace from poor heredity, it may itself add heavily to the debit side. With the very best of health backgrounds, environment may damage body and mind beyond repair. Under environment we include everything that touches life from without—people, things, work, play, home, school, social life, business life, college-life, etc. Among factors of environment damaging to mental health are overemotional family life, overstrict home discipline or the lack of needed discipline; overfeeding, underfeeding, wrong diet, lack of proper exercise, stimulants, drugs, overstimulation, overprotection, too much hardship and privation, loneliness, poor educational methods, immorality, etc.

Personal Reactions

What will decide whether a human being can resist, successfully, bad tendencies in heredity, or in environment, or in both, and keep a reasonably balanced mind? It demands insight, ambition, will; and if these remain the body can be forced to saving ways of health, and body and mind can largely make their own environment. But with heavy handicaps of heredity or environment, or both, and poor insight, or lack of desire, or weak will, nothing can save the mind from neurotic taint or worse—nothing but obedience to some one strong enough to control the habits of that life, until self-control is born. And there is a hope that it can be born in the most neurotic or neurasthenic, so long as the mind is sane.

But after all, a large number of people whose mental processes are not normal, have only themselves, their poor emotions, their lazy wills, their hazy thinking to blame. We except what are called the heredity insanities—dementia præcox and the other dementias and the manic-depressive groups and paranoia and psychasthenia—for in these cases, possibly with the exception of the manic depressives, even the most perfect environment could probably not prevent the disorder from asserting itself. Many neurotics, neurasthenics, and hysterics are curable if they will seriously undertake to fulfil the laws of physical and mental health—simple laws, but ones which demand a strengthened will to carry out.

CHAPTER X
ATTENTION THE ROOT OF DISEASE OR HEALTH ATTITUDE

The Attention of Interest

Attention naturally follows interest. It can, however, be held by will to the unappealing, with the usual result of transforming it into a thing of interest.

One of the laws of the mind we have already stressed is that what we attend to largely determines what we are, or shall be. The interests which secure our consideration may be the passive result of emotional life, the things which naturally appeal, which give us sensations that the mind normally heeds; or they may be the active result of our will which has forced application upon the things which reason advised as worth acquiring.

We found that the beginning of health of mind consists in the directing of thought toward the health-bringing attitude. We have seen how quickly the normal mind can be diverted from the undesirable by a new or stronger emotional stimulus. We found that the sole appeal to attention in the baby-life is through the emotions, and that it is natural throughout life for the mind to heed and follow the interesting; which is only another way of saying that thinking follows where emotion leads, unless volition steps in to prevent. The supreme test of the will's power is its ability to hold the train of thought in the line that reason directs, when feeling would draw it elsewhere. This ability marks the man who does big things; while the inability to ever turn attention away from the interests proposed by feeling assures weakness.

Some of the most charming people we shall ever know are those temperamental children of happiness whose interests are naturally wholesome and externalized, whose natures are spontaneous and joyous, and who live as they feel, seemingly never knowing the stress of forced concentration. With them attention follows feeling, feeling is sweet and true, and volition simply carries out what feeling dictates. And life may not be complicated.

But there is another class whose attention also follows in the ways of least resistance; and life for them is a wallowing in the morbid and unwholesome. In them feeling is perverted, they seem to see life habitually through dark glasses; they passively attend to the sad, the distressing, sometimes the gruesome and the horrible with a sort of pallid joy in their own discolored images. The first group puts joy in all they see, because they are brimming full of joy themselves. These others find only the unwholesome in life because their minds are storehouses of it. We say that each type has projected himself, that is, has thrust himself out into the external world, and is standing back, looking at his own nature and calling that the universe.

But neither of these two groups can long withstand the stress of a world they only feel and have never attempted to comprehend. The irresponsibly happy ones are too often crushed and broken when life proves to bring loss and failure and disappointment; the morbid probably will cease some day to enjoy their melancholic moods, and be unable to find their way out of them. If both had learned to control attention, they might have been saved. The happy, care-free child of the light is at desperate loss when the sun he loves is obscured, if he has not learned to look upon the far side of the clouds to find that there they glow golden with the rays temporarily shut from him. Because clouds were not interesting to him he never attended to them—and now he cannot. If the pessimistic, morbid one had looked away from the shadow to the sun it hid he, too, in the end might have seen with sane eyes and lived so wholesomely as to find all the good there was in life. Willed attention, rather than spineless feeling distractibility, might have saved him.

When thinking can be forced to follow where trained reason directs, and can be kept in that direction, the greatest problem of physical and nervous well being is solved. To the nurse there is no other principle of psychology so important. But no child ever had his attention diverted by reasoning alone. The object at which you wish him to look must be made more impelling than the one he already sees, or he must want much to please you, else he only with his eyes will follow your command while his mind returns to his real interest; and the second you cease to command that eye service, he looks back to the thing that was holding him before. The beginning of all education is in arousing a want to know; in turning desire in the direction of knowledge.

I am an undisciplined child and I want only candy for my lunch. It is not good for me. Milk is what I should have. I don't want it. You may deprive me of the candy and force me to drink the milk, and I can do nothing but submit. But I rebel within, and I am only more convinced that I "hate" it and want candy, and that you are my natural enemy because you force the one upon me and deprive me of the other. If I were insane and so, of course, could not be reasoned with, this might be inevitable. But it would be unfortunate. In that case, if possible, do not let me see the candy; let only the food it is best for me to have be put before me, and perhaps eventually I shall come to want the more wholesome thing—for it is better than the hunger.

But as it happens I am a perfectly normal person, only I am sick. I am tired of bed, and want to sit up—and it does seem that I should have my desire. The nurse, wise in her knowledge of sick "grown-ups," who are, after all, very like children, will find a way to divert my mind from the immediate "I want" to something which I also can be led to want. I may agree that I want more the better feeling an hour from now. Perhaps her humorous picture of the effects of too early freedom on my condition, or of my body's urgent demand for rest, regardless of my mind's wish; perhaps only a joke which diverts me; perchance the "take-for-granted you want to help us out" air; mayhap the story to be read or told; or simply the poise and quiet assurance of the nurse who never questions my reasonableness and acquiescence; perhaps her confidence that this will serve as a means to the end I covet—will result in my gladly taking her advice, and my perfect willingness to wait for new orders, while I indulge in beautiful plans I shall carry out when they finally arrive.

In other words, with the sick as with children, attention naturally follows interest. And the good nurse realizes that it is not wise to force co-operation when she can secure it by diverting her patient's thoughts to another interest than the one now holding him. Very often, merely by chatting quietly about something she has learned has an appeal, she can make the patient forget his weariness and boredom, or his resistance to details of treatment. The very milk he is refusing to drink may be down before he realizes it. But right here lies a hidden reef which may cause wreckage in the future. It is good therapy to divert attention by appealing to another interest when the patient is too sick or too stubborn or not clear enough mentally to be reasoned with. But if this becomes a principle, and his reason and active co-operation are never secured to make him choose the way of health for himself, the hour he is out of the nurse's hands he reverts to the things that now happen to appeal to him. Then unless some wise friend is near to continue her method of making the reasonable interesting, the advice of reason can "go to smash."

There has been a very constant illustration throughout the past of the unwisdom of relying upon diverted attention alone as an effective therapeutic agent. We hope this will not illustrate our point so clearly in the future. The drunkard, who is just recovering from a big spree, and feels sick and disgusted with himself, and sore and ashamed, is appealed to in glowing terms of the wellness and strength and buoyancy of the man who never drinks. He has no "mornings after." The Lord is just waiting to save this dejected victim of alcohol from his hateful enemy who has made him what he is at this hour, and will forgive all his sottishness, his sins. He will be respected; he can command the love of his family again. He will no longer be a slave, but a free man. Right now, respect of the world and love of family and friends, and cleanness, and the forgiveness of a good God are infinitely more interesting than this splitting headache, this horrible sick feeling. And attention may be very readily diverted. This promised new life is more attractive than the present. It is easy to keep attention there. And he reforms. He swears off "for keeps." He is a happy man, a free man. For a few days or weeks, perhaps even longer, he glories in his new self-respect. It is a strange and enticing sensation. Then one day something goes wrong. He loses some money, or he is awfully tired, or the wife and children bore him, and all of a sudden the one greatest interest in the world is a drink. And because his thinking can always be led by his feeling; because he has never learned to force it to go elsewhere, he has his drink. Appealing to his emotions did not and cannot save him unless that appeal is followed at the right moment by awakened reason, which will look at the whole proposition when the mind is at its normal best, and choose to follow where rational feeling directs. Nor will reason save unless volition comes to its support and strongly backs it up and enforces what it advises.

The Attention of Reason and Will

So the good nurse will not consider her work done when she has diverted mental processes into channels of co-operation. When the patient, who is capable of reasoning, knows the why of his treatment, and realizes that he can only keep well as he himself takes over the job and puts his mind on things out side of his feelings, and carries out the doctor's instructions for the sake of securing a certain end—then he has been under a good nurse. This wise helper never "preaches," but makes the healthy goal very desirable, stirs up an ambition to attain it, and prods the will to keep on after it despite anything feeling may say.

This attitude on the part of the nurse presupposes that her own attention, while with her patient, is upon him and upon securing his health, and not upon her tiredness, or boredom, or headache, or the party tonight, or the man who has asked her to go to the theater with him tomorrow. She, surely, must learn to direct her thoughts where reason suggests, and to gain new interests through willed attention, or as a nurse she is less than second rate. Nor can she get the best results until she can turn with a single mind to the patient at hand as the immediate problem to be solved. And probably neither nurse nor doctor does any better service, except in saving life itself, than in keeping the patient from thinking constantly of himself and his ills. For it seems of little use to have made some people physically well, if they are to carry through prolonged years the curse of constant self-attention, self-centeredness, an ingrowing ego.

There are a few simple laws of the mind hinging upon attention which are today being impressed upon teachers in every department, in kindergarten, public school, college, and university. And they are as necessary to the nurse as to the teacher. Three of them we have already discussed:

1. Attention naturally follows interest.
2. Attention may be held by will where reason directs.
3. New interests grow out of willed attention.

A fourth we shall stress before considering the use the nurse can make of them:

4. The thing to which our chief attention is given becomes the most important thing.

Do not contradict this too quickly. Don't say that nursing gets your chief consideration because it is, of necessity, your profession; but that you love your music infinitely more, and look forward to that through all your hours on duty. If this merely proves that music is distracting your attention, you are doing your nursing as a means, and not as an end; you give it probably all the attention necessary for good work, but your real desire is music. Your chief attention is directed toward that goal. Hence music is to you the most important thing. If your will is sufficiently trained to keep you from consciously thinking of it, still you are dreaming of it and working for it. You may make a very good nurse, but you will never be as excellent a one as the woman from whom nursing demands first and chief attention.

We sometimes speak of one woman as a born nurse, and say of another, "She's a good nurse, thoroughly conscientious, but not a natural one like Miss X." It only means that Miss X's main purpose in life has always been caring for the sick, while Miss Y's secondary concern is that. There is a third, however, who may be sidetracked into nursing, but whose chiefest interest and attention in life has not been so much a certain profession or accomplishment, but a passion for people, with an ability to enter into their lives understandingly. She may not care for nursing in itself. It is only accidental that her thoughts were turned to it. But her liking for people makes it easier for her to concentrate attention on the details of nursing, as thereby she is fulfilling her life's ambition in studying and serving human beings. She may be a real success if she can only convince herself that this is her forte. If not, and she dreams of other fields of service, her concentration on the thing at hand is not perfect enough for her to compete successfully with the "born nurse."

Whatever it is, the thing that gets our chief attention is the most important to us. It may be lack of appetite, or pain in the side, indigestion, general disability, discomfort, the mistreatment we once received, the mistake we once made, or the sin we committed—whatever it is that holds our attention, it is the most absorbing and interesting thing in the universe, though it may be an utterly morbid interest, an unhappy attention. But it blots out for the time the rest of the world. A big hint for the nurse exists therein. Let her try in every lawful way to divert her patient's attention from the disease-breeding stimuli toward the happy and wholesome ones.

For the nurse herself in the care of patients let us draw some conclusions from these laws of the mind's working:

1. Have a goal in view for the patient's health of both body and mind.
2. Work toward instilling in your patient a health ambition—a pride in health.
3. Remember that overcrowding the mind defeats your purpose of making one clear impression.
4. Win interest by any legitimate means to the next step toward the goal, and only the next.
5. Work for attention to hopeful, courageous, and happy things.

Let us as nurses remember always that it is for the patient's sake and not for our own that certain results must be obtained. Our work is usually in helping the doctor to get the best possibilities out of the material at hand, and we cannot hope to change the fabric. But we can help to repair it; we can sometimes influence the color and suggest some details of the pattern, or assist in the "making over" process; and when the fabric is substantial and beautiful we may assist in preventing its marring. So we may help to evolve a body-health and mind-health attitude from what seemed the wreckage of a disease-accepting mind; or we may have the great privilege of warding off the disease-accepting attitude. But always, in all our care of patients, let us not neglect or fail to use wisely this central fact of psychology; that anything that gains attention, even for a moment, leaves its impress on the mind; that the direction of attention determines our general reaction to life.

CHAPTER XI
GETTING THE PATIENT'S POINT OF VIEW

What Determines the Point of View

The point of view of any individual depends upon temperament, present conditions—mental and physical—and the aim of the life. That is, it depends upon his inherited tendencies plus a unique personal something, plus all the facts of his environment and experience, plus what he lives for.

Richard and Jim both live in Philadelphia, Richard on Walnut Street and Jim on Sansom Street. Richard's father is of the best Quaker stock, with hundreds of years of gentle and aristocratic ancestry behind him. He followed his father and his grandfather into the profession of medicine, and is a well-known specialist, alert, keen, expert, and deservedly honored. He is at home in Greek and Latin, French, and the sciences. He selects at a glance only the conservative best in art and music and literature. His world is a gentleman's world, a scholar's world, and the world of a scientist and a humanitarian. And Richard, his son, is true to type.

Jim's father is the ash man. His world is in the alleys and basements. His pastime, cheap movies, and the park on Sundays. When he is not working he is too "dead tired" for anything heavier than the Sunday Supplement or perhaps the socialist club-rooms, where he talks about the down-trodden working man and learns to hate the "idle" rich. He spends his money on food and cheap shows and showy clothes. He talks loudly, eats ravenously, works hard, is honest, and wants something better for his children than he and the "old woman" have had. His music is the street-organ, the movie piano, and the band—some of it excellent too—but none of your dreamy stuff—good and lively. And his son, Jim, is true to type.

After the Armistice Jim and Richard, who have fought for months side by side, go to Paris together. Richard may "have a fling" at Jim's amusements for the sake of playing the game and "seeing how the other half lives" and all that—but before long we shall find him in the high-class theaters and restaurants, visiting the wonderful art collections and libraries, riding in luxurious automobiles, and staying in the best hotels he can find. And even though Jim may have saved Richard's life and Richard is eternally grateful, and loves Jim as a "dandy good scout," their ways will inevitably drift apart when the one big common interest of fighting together for a free world is over. They will always remember each other. Jim will decide that a "highbrow" can be a real man, and Richard will ever after have a fellow-feeling for the "other half" and think of them now as "folks." But Jim is not at home in Richard's neighborhood and circle; and Richard is a fish out of water in Jim's. The point of view of each has been largely determined by his heredity and his environment.

But suppose Jim isn't true to type. From the time he was a mere youngster the ash-man life did not appeal to him. In school he liked the highbrow crowd; he "took to" Latin and literature. He has a feeling of vague disgust when he sees a vulgar picture, a shudder when the street-organ grinds. There is something in Jim different. He isn't in tune with either his immediate heredity or his environment. The contribution from some remote ancestor has overbalanced the rest, and Jim becomes a professional man.

Or perhaps Richard breaks his father's heart. Instead of following the trail already made, he cuts loose, frequents vulgar resorts, hates his school work, becomes a loafer and a bum—and, finally, a second-rate day laborer. Again, what he is himself, his "vital spark" has been stronger than immediate heredity and environment, and has broken through.

Getting the Other Man's Point of View

Our points of view are very frequently merely hereditary or acquired prejudices, hence altogether emotional rather than rational. We only with great difficulty see things through another man's eyes. It necessitates comprehending his background fully, and standing exactly where he stands, so mind and eyes can both look out from the same conditions that confront him. And this is only possible for the man or woman possessed of a vicarious imagination. Such an imagination, however, can be cultivated.

You hate my father. He injured yours—unjustly, to your mind, of course, for yours can do no wrong. From my point of view this father of mine is a great, good man. From your point of view he is wicked and cruel. We are both honest in our emotion-directed opinions. Until you can know my father as I know him, and I can know yours as you know him, we shall never agree about them. But I can learn to understand why you feel as you do, and you can learn to understand why I feel as I do. I can put myself, in imagination, in your place, and see that other man as my father, and pretty well grasp your point of view, and you can likewise get mine.

After all, the law is very simple. Each man is the result of the things he puts his attention chiefly upon; and he puts it naturally upon the things which his forebears and his surroundings have held before him. The rare person and the trained person can assert the "vital spark" of his own personality and tear attention away from the easy direction and force, and hold it somewhere else. So he can change his points of view by learning that there are other vantage grounds which direct to better results. With some one else to lead the way and give a bit of help, or with the urge of desire to understand the new viewpoint, or by the drive of his will, he can change his own.

Let us not forget that what we see depends on whether or not our eyes are normal, on where we look, or on what kind of spectacles we wear. Two things we can change—where we look, and the spectacles. If our eyes were made wrong we probably cannot change that, but we can often correct poor vision by right artificial lenses. There are people doomed to live in most unattractive, crowded surroundings who make a flower-garden of charm and sweetness there, or, without grounds, keep a window-box of fragrance. The normal person can pretty largely either make the most impossible environment serve his ends or get into a better one. So we can usually look to something constructive, helpful, attractive, or beautiful; and we can refuse to wear blue spectacles.

We nurses soon realize that there are just about as many points of view as there are people, and that if we would help cure attitudes as well as bodies, and so lessen the tendency to sickness, it behooves us to learn to see what the other man sees through his eyes or by the use of his glasses, from where he stands.

Let us try just a few experiments. Hold your pain and suffering from your appendix operation, and disappointment because you can't be bridesmaid at your chum's wedding, up close to your eyes, and you cannot see anything else. They crowd the whole field of vision. Look at the world from the eyes of a spoiled woman of wealth who for twenty years has had husband, friends, and servants obedient to her every whim. She has grown selfish and demanding. What she has asked for, hitherto, has been immediately forthcoming. Now she is ill, and she naturally considers the doctors and nurses mere agents to secure her relief from discomfort. She is willing to pay any price for that—and still she is allowed to suffer. From her point of view it is utterly unreasonable, inexcusable. What are hospitals and nurses for, anyway? And she is carping, critical, and disagreeable. Her attitude is as sick as her body. How could it be otherwise?

Look about you from an aching mind and body, after days of suffering and sleeplessness, and unless you are a rare person and have a soul that sees the sunshine back of everything—you will find the world a place of torture. Look out from despair and loss of the ones you love best, or from failure of will to meet disaster, and everybody may be involved in bringing about your suffering, or in effecting your disgrace.

Look out on the world from the eyes of the immigrant who has lost all his illusions of the land where dollars grow on the street and where everyone has an equal chance to be president, and if you do not cringe in abject humility, you are not unlikely to be insufferably self-asserting, considering that the world has robbed you and that now it is your turn to get all that is coming to you. So you make loud demands in a rude, ordering voice. The nurse is there to wait upon you—and finally you will have your innings.

Look out from the resentful eyes and smarting mind of the negro who is just beginning in a northern city to realize that his boasted "equality" is a farce, and you will try to prove to the white nurse that you are as good as anybody. You are impossible; but back of all your bravado and swagger and rudeness and complaint of neglect because of your color, you realize that you cannot measure up. You know you belong to a different race, most of whose members are daily giving evidences of inferiority; and you are sure that the nurse is thinking that.

Look from the eyes of the "new rich," or the very economical, and you are going to get your money's worth out of your nurses.

The nurse who can get back of her patient's forehead and put her mind there and let it work from the patient's point of view, will learn a saving sense of humor, will be strict without antagonizing, will clear away a lot of mental clouds and help to make permanent the cure the treatment brings.

One can often judge very truly a patient's real character by his reaction to his sickness. On the other hand, frequently it only indicates that he has not yet properly adapted himself to a new experience and a trying one. We hear so often, "Why, she's a different person these days, since she's feeling better. It's a joy to do things for her." She was the same person a while back, but had not learned to accept discomfort. Any of the following list of adjectives we hear applied to our patient again and again by the nurses:

unreasonable	stubborn	lazy
cranky	resistive	unco-operative
hipped	obsessed	hypocritical
of mean disposition	excitable	fearful exacting
dissatisfied	deluded	will-less
undecided	wilful	self-centered
morbid	doubtful	demanding
retarded	abusive	depressed
spineless	self-satisfied	

Unpleasant terms they are, and condemning ones if accepted as final. When the nurse realizes that under the same conditions she would probably merit them herself, she becomes more anxious to remove the conditions, and less bent upon blame.

We must admit that the highest type person, when sick of any physical illness, does not deserve such descriptive terms as these. But they are the rare folks, few and far between; while the great mass of us have not acquired more than enough self-control and thoughtfulness for the ordinary routine of life. We are weakly upset by the unexpected. If it is a pleasant unex pected, we are plus in our enthusiasm, and people applaud; if the unpleasant unexpected, we fall short, and people deplore our weakness. If we learn our lesson of self-control and adaptability, and gain in beauty of character through experience, it has served a purpose. But the nurse deals with the average of human nature, and she finds their reaction faulty. Very often, if she is observant, she will discover that a patient responds in a very different way to some other nurse, who somehow finds that "trying" sick woman charming or thoughtful, likable or sweet. Of course, it may be because the other nurse weakens discipline and caters to the patient's whims; but it is just as likely to be because she has tempered her care and her strictness with understanding. She has grasped the patient's point of view; and with that start, the chances are 50 per cent. more in favor of the patient grasping and acceding to the wise nurse's point of view.

Shall we not remember that our trying, cranky, stubborn patient is a sick person, and learn to treat that stubbornness or crankiness as a symptom indicating her need, just as we would a rising temperature?

When we can meet her attitude with comprehension, and, if necessary, with quietly firm disregard, then we are beginning to be good nurses.

Some of the most common of these sick reactions with which the nurse must deal are enhanced suggestibility, repression, oversensitiveness, stubbornness, fear, de pression, and irritability. And each one demands a different method of approach if real help is to be given.

Old Isaac Walton wrote a book many, many years ago called "The Complete Angler." He was a famous amateur fisherman, and he says there are only three rules to be observed and they will bring sure success:

1. Study your fish.
2. Study your fish.
3. Study your fish.

If the angler follows these directions, he is not apt to offer the wrong bait. When he knows all their little peculiarities, he will know how to catch his fish. The "complete angler" has an unlimited patience and an infinite sense of repose and calm. He never hurries the fish, lest they become suspicious of his bait. And he proves that these three rules work.

The nurse who accepts every patient as like every other, and treats him accordingly, will never be a great success. The nurse who "studies her fish" and learns their psychology, will be a therapeutic force. She will know the why of the way that patient acts.

The Deluded Patient

If the patient's mind is temporarily clouded through infection or suffering, he may be reacting to a delusion, an obsession, a fixed idea of disability, a terrifying fear. Sometimes he persistently refuses food, and gives no reason for it. The unthinking nurse is tried, puzzled, and irritated. In other ways, perhaps, the patient seems quite normal. But, after all, the explanation is very simple. He probably is as confident that the food is poisoned as you are that it is as it should be. No arguing would convince him, for, to his mind, the nurse is either a complete dupe or an agent of the people whom he knows are plotting his death. And urging him only strengthens his conviction.

The writer recalls one such case of a patient who had to be tube fed through many months, though a tray was set before her three times a day—and as regularly refused. Then one day she was seen slipping food from off another patient's tray and eating it greedily, not knowing she was observed. When questioned, though she had never before given a reason for refusing food served to her, she said that "they" had nothing against Mrs. B., so wouldn't try to poison her. Her reasoning was excellent when one accepted her premises. She had bitter enemies. They were not enemies of Mrs. B. and would not harm Mrs. B. Therefore she dare not touch her own food, but could eat Mrs. B.'s if no one knew.

These deluded patients live in a world we often do not sense, a world whose reality we do not appreciate. The nurse, after much experience, finds that there is a key to every resistance, to every lack of co-operation, to abnormal attitudes and actions. She realizes that a powerful emotion of desire or fear, of love or hate, of ambition or self-depreciation, of hope or despair, of faith or distrust, unchecked by reason or judgment through the years, has provided a soil upon which emotional thinking alone can grow. The patient is a mere puppet of the suggestions of emotions which may not be at all pertinent to the facts.

Nursing the Deluded Patient

The nurse soon realizes the uselessness of attempting to argue a patient out of his delusions, of trying to convince him that the things he sees and hears and perhaps tastes and feels, are but hallucinations. Her very insistence only fastens his attention more firmly upon the false conclusion or makes him more convinced that his mind is giving him a true report from the senses of sight and hearing and taste and feeling. But often a quiet disregard of the delusions while the nurse goes on her way and holds her patient to his routine, consistently and confidently, as she would in case they were not true, will eventually cause him to question their reality just because no calamity results. The nurse acts as if these delusions and hallucinations were non-existent in reality, and when the occasion arises, through the patient's questioning, she urges him to exert his will to act also as if they were not true; to try it and see what happens. Arguing, also, she finds, usually antagonizes or makes the patient stubborn. He cannot prove by her logic his point, but he "knows" from inner experience that he sees what he sees, hears what he hears, and knows what he knows. The fact that the nurse does not is merely annoying evidence that she is blind, deaf, or stupid to these things of his reality. He knows he is lost and damned, or tainted; that he is King George, Cæsar, or the Lord, as the case may be; or that his internal organs are all wrong. He "feels" it and the nurse can't—therefore, he alone has true knowledge of it. In the end, the wise nurse who never disputes with him, but leads him on to action which utterly disregards these things, may bring about a gradual conviction in the patient's mind that a man couldn't do what he does if all these things were true; and the delusion slowly may lose its force or the hallucination fade away. Many patients drop them from their lives entirely. Many others in whom dementia is not indicated, or in whose cases it is indefinitely delayed, can come to an intellectual realization that all these things are fantasies, and do not represent reality; that despite their continued, frequent, or occasional demands upon feeling life, they can be consistently ignored. These psychopathic individuals may act as they would if the delusions never came henceforth to their consciousness, and so be enabled to live a comparatively normal life.

The Obsessed Patient

A patient who is suffering from obsessions must carry out certain abnormal actions, or be wretched. She cannot do otherwise. It is as though she were forced by some outside agent, though the forcing is actually from within. When the nurse realizes this, and the more essential fact—that many patients, who have not true obsessions, yet have a tendency toward obsessed ways of thinking and doing—when she comprehends it almost as she would if she were the victim, then she is ready to help the patient by gently making the action impossible, and at the same time diverting attention.

The Mind a Prey to False Associations

Sometimes a nurse reminds a patient of some one in the past who has complicated her life in an unhappy way, so she distrusts or dreads her or is made constantly uncomfortable in her presence. In such a case, if the nurse reports her patient as resistive, or fearful or cringing, or distrustful, she is really misrepresenting her; for under another's care that patient may show an entirely opposite reaction.

The nurse can only sense the strength of the influence of heredity and environment and habit of thought, which would give the explanation of many things in her patient's attitude. Nor can she realize just what shade of meaning certain phrases and words have for her charge. To the nervously overwrought person the most innocent reference—father, sister, wife, home—may bring concepts that are unbearable. The association of the word may make for deep unhappiness, of which the nurse knows nothing. But she can learn that all these things do influence attitude, can appreciate the difficulty of her patient's effort at adjustment, and do all in her power to make that adjustment possible. If the patient is reasonable she can appeal to her reason. If she is too sick for that, the nurse can use happy suggestions. If the mind is deluded and obsessed she can use firm kindness. She can learn what loss of privileges will affect the rude and unco-operative patient, and may be allowed to try that. She can sometimes help the patient to self-control by making her realize that after each outburst she will be constructively ignored.

But the point we wish to make is this: There are some sick reactions which the nurse, if she recognizes as such, can help the patient to transform into wholesome ones. At the very least the wise nurse can learn to simplify her own difficulties by accepting the unpleasant patient as possibly the result of her illness, and refusing to allow her trying attitude to get on her nerves. The patient may be reacting normally to the stimulus her untrained and toxic brain received. And when the nurse can see into the other's mental workings, get her point of view, she is ready to give fundamental help.

CHAPTER XII
THE PSYCHOLOGY OF THE NURSE

The mind can be as definitely developed and strengthened as the body. The man who has suffered for years an organic disease will never have the same force as he who has never been seriously ill; but his constitution can be built up and made as efficient as possible within its limitations. Many a man or woman who has an organic heart disorder, through treatment and the proper exercises gradually increased, can very often approximate through many years the output of a normally strong person. The individual weakened by a tuberculous infection can frequently, by following a prescribed regimen for a time, by wise, scientific diet and rest treatment and the help of the out-of-doors, then by carefully increased physical activity, finally live the useful, average life. But it takes scientific care to evolve the weak body into a strong one; and in some cases, at best, it can never stand the same strain that the uninjured one carries with ease. However, even damaged bodies can be made very productive within their limited spheres. Also the naturally perfect physique can quickly become unfit through neglect or infections or misuse.

In the same way, and just as definitely, can the mind be developed and strengthened. Some are by nature keen, alert, brilliant. They may develop into masterfulness; or they, too, may degenerate, through abuse, or from the effect of body infections, into uselessness. The germ-plasm has foreordained some individuals to psychic disorders; but training and mode of life can modify many of these defects. And the average mind, like the average physical organs, can be made more efficient through partaking of the proper mental food, through careful training and wise use.

No more urgent necessity faces the professional woman than this of training her mind to its highest productiveness. Argument is not needed to convince intelligent people today that the accomplishment of life depends upon mentality.

Let us look into the very A, B, C's of mind development, and as nurses undertake to equip ourselves to master our profession from the ground up. The first essential is ability to think clearly.

Steps to Clear Thinking:
1. Accurate perception, with attention to the thing that reason chooses.
2. Association of ideas.
3. Concentration, acquired by the help of emotion and will.
4. Emotional equilibrium, which refuses to allow feeling to obscure judgment by leading reason astray.
5. Self-correction.
6. Automatic habits, which free the mind of all unnecessary crowding.

Accuracy of Perception

The beginning of learning is perception. Keen, accurate perception at the time of first introduction of a new fact or thought, and the linking up of that new material with something already in consciousness, insures in the normal mind the ability to remember and use that fact or thought again. The things casually perceived and not definitely tied up with something else are soon forgotten by the conscious mind.

You pass a florist shop where a score of different flowers and plants are displayed. If your thoughts are intently on your errand you may glance in, see flowers, color, perhaps a riot of colors only—and beauty; and you feel a glow of pleasure from the sight. But a moment later you cannot name the blooms in the window. Perhaps roses come to mind because you have very special feeling for them;

or carnations, or sweet peas. But the window as a whole you perceive only as flowers, and color, and beauty. You cannot describe it in detail, for you gave it only passive attention.

But if you went to that window to know its contents; to find out what the florist had in his shop, because you are very interested in all flowers and plants, then you can tell minutely what is there. You had a purpose in perceiving the window; your will held attention upon each object in turn; and your love of flowers (an emotion) eased the effort of volition when it might have tired.

Perception, then, is of three kinds: passive, incited by interest, and directed by will. And the perception which is the basis of accurate knowledge is one of keen interest, or of will, or of interest plus will.

Training Perception

The nurse who demands of herself that she perceive accurately paves the way for accurate, deft service in her profession. There are constant means at hand for training in the art. Suppose you try to get so definite a picture of each ward or room you enter, in a swift but attentive examination of its furnishings and their locations, and of the patients, that you can reproduce it to yourself or a friend some days later.

You come into a large ward, with a row of beds on either side of the door, and a wide central space between. How many beds in each row? There is a table at the far end of the room, opposite the door, and a nurse in white is writing there. Why does she wear white? What is her name? To your right is a closet-like room opening from the ward. That is a medicine-room, you are told. How many windows has the ward? You glance from bed to bed with a rapid passing in review of the patients. Which ones seem to you very ill? There is a large white screen about one. You are told that when treatments are given the screen is put there, or that when a patient is dying the bed is screened. You look for the ventilators, and see how many are open and how they work. You see a room-thermometer, and ask at what temperature it is kept. The nurse explains that a certain degree is ordered, and that, so far as possible, the ventilators are operated to insure that.

If your attention has followed all these details with careful, accurate perception; if you have grasped them clearly, one by one, at the time, you will be able to answer quickly next day when some one asks how many patients the wards accommodate, and how many beds are vacant. You can describe the lighting and ventilation, the room temperature, etc. And later on you will quickly see to it that a screen is properly placed when you know treatments are to be given.

Association of Ideas

After the first few years of life practically nothing enters consciousness that cannot by some likeness or contrast or kinship be connected with something already there. Were it not for this saving economy memory would be helpless. So the nurse who is in earnest and eager to master her new work will not only perceive carefully each detail of arrangement, but in two or three days at most will know each patient there; she will have worked out a system of associations, remembering not a meaningless name, but an individual with certain characteristics which she ties up with her name, and so gives it a definite personality. She thereafter recalls not merely a patient, but a very special patient; and as she comes to mind she brings a title with her, which is her symbol. Likewise when her name is spoken or thought, she herself comes into the nurse's immediate consciousness. A bed in a certain part of the room will be no longer merely a bed, but Mrs. Brown's bed. Remembering can be made easy by using some such method as this:

The first bed to the right as you enter is Mrs. Meade's. She is the woman with the broken hip. The next is Mrs. Blake's, that blonde, big woman who wants more attention than any one else. The third is Mrs. Bunting's. She has wonderful, curling black hair, and a nice response to everything done for her. The next beyond is Mrs. O'Neil's. She looks as Irish as her name sounds, and you will remember her by that. So each bed comes to mean a certain patient, and each patient comes to suggest the ones on either side of her — her neighbors. Blondeness and bigness together call Mrs. Blake to mind. Broken hip means Mrs. Meade, etc. Each individual on that side of the ward becomes associated with a name which stands for definite characteristics.

Then you begin at the left bed nearest the door and follow the occupants back on that side. You may re member better by jotting them down in order of the beds, with names and a brief comment on each patient. Keep that list on a small card in your pocket for reference for a day or two, then depend on memory entirely. I have personally found this an excellent method.

You are expected to be able to turn quickly to any medicines needed in emergency, and you soon learn to remember them and where they are placed by the arrangement into classes or kinds, which most hospitals require. Cathartics are together, hypnotics together, etc. So when you want cascara you associate it with cathartic and turn to that shelf. You learn very soon that poison medicines are kept apart from the others, and quickly associate the poison label with danger to patients, necessity of locking safely away and hiding the key from any but those responsible for the care of the sick.

Learning to look closely at the patient's face, instead of casually glancing at her when you care for her, makes it possible for you to note changes of expression, heightened color, dilated pupils, a trace of strain, etc. Then try to find the exact word that will express what you see. Such experiments in perception and attention, association and memory, repeatedly demanded of yourself — i. e., the being able to recall and describe in detail the room- or ward-arrangements and to place the patients accurately, as we have just described — will prove invaluable practice, helping you to attend to every change in your patient's demeanor and expression, which may prove significant symptoms. And remember that while the mind can only contain so many isolated facts, yet there is no limit to its possibilities when the power of association of ideas is employed.

Your first step to clear thinking is accuracy of perception, with attention to the thing reason chooses; your second is association of the things perceived, a grouping of them to fit in with each other, and with what is already in the mind. And both imply the third—concentration, aided by emotion and will. For passive attention and haphazard associations assure the opposite of clear thinking.

Concentration

How to Study.—You learn sooner or later from experience that the quickest and best way to learn anything new is to give it your undivided attention at the moment; to perceive one thing at a time and to perceive it as something that is definite, or as some quality that is unblurred. One of you will spend three hours on an anatomy lesson, another two hours, while a third nurse may give it a half-hour of concentrated study and know it better than either of you, if you have been day-dreaming, or talking, or rebelling at the "luck" which keeps you indoors learning about bones, when the tennis-court is so inviting. True, some minds have better natural equipment and some have better previous training than others. But the average mind could learn a lesson well in much less time than is spent upon learning it poorly. Few people hold their attention strictly to the task at hand if something more interesting beckons, or if they feel tired, or "blue." But you can learn to do it.

Put aside a certain amount of time today for study; hold your undivided attention on your lesson, regardless of how many pleasanter things appeal. When your eyes or your thoughts wander from your note-book, bring them back forcibly, if need be. Your first task is to keep your eyes there, instead of letting them follow your roommate's movements, or resting them by watching the street below. But it is easier to do this than to make your mind grasp the meaning of the things you see. You may read two or three pages, and not receive one idea, not even be able to recall any words from the context. Your eyes are obeying your will and seeing the words, but your mind is "wool-gathering." Now take yourself in hand firmly. If you are really a bit fagged, try some deep-breathing exercises before the open window, bathe your face in cold water. Then read a paragraph, close your book, and write, if you are not alone, or repeat to yourself aloud, if your roommate is out, what that paragraph says—its meaning. If you cannot do it, read it again with that end in view. Repeat the process, and hold yourself to it day after day, if necessary, until finally will has won the battle, or, better still, your will to learn has been reinforced by an interest in the very competition with yourself, if not yet in the contest. Then, as you learn some facts from your notes, use your imagination to apply them in real life.

The triceps muscle. What is it for? Your notes inform you, and then it is really interesting to see how it performs its function. What origins and attachments must the triceps have to make it extend the arm? Your notes say that a muscle tends to draw the part to which it is attached toward its origin. This triceps muscle straightens the arm. In that case it must oppose the flexion at the elbow. How is that likely to be done? The triceps must start somewhere above the elbow, and quite far above, too, to be able to make a straight angle of an acute one; it must start toward the back in order to draw back the forearm; and be attached to the back of the bone below. Also it must be quite a long muscle. So much reason tells you. Now let me see how it is done, in fact. And you find that the triceps has three origins high above its one attachment as a tendon, to give it a good strong pull. These are in the outside of the humerus and in the scapula. That is logical, and you will remember it.

Now how does the arm bend? What pulls against the triceps? And you are interested before you know it.

There is nothing, good, bad, or indifferent, but has some points of interest if the mind turns its entire attention to it. But our tendency is to grow tired of calling back our wandering thoughts again and again to the thing that is hard, dry, or stupid. And we need more incentive than just the doing of the duty because it is to be done. We need a compelling interest in the goal to encourage our wills to concentration on the less interesting. Let us first think out the why of knowing anatomy if we are to be nurses. And if the profession of nursing is the goal, let anatomy become just the next stretch of the road that leads to it.

Concentration can be acquired. It may require three hours at first to learn your lesson; but later on you will do it in two, then in one, and perhaps in less. And when you can sit down with your notes and learn them with voices about you—perhaps; with some one else in the room; with a party an hour ahead; when you can disregard all but the work at hand, then you can concentrate, and the big battle of your life as a student is won. Study is no longer drudgery. Lessons occupy much less of your time and leave you more free hours. Because you give them your whole mind you learn them in a fraction of the hours hitherto wasted upon them, when you studied with divided attention. When you are doing clear thinking on the thing at hand, satisfactory results are assured.

Self-training in Memory

Hand in hand with clear thinking goes reliable memory. But so many of us have it not, and feel its need so strongly that we shall consider for a moment some means of training it.

William James holds that brain-paths cannot be deepened; that memory is not strengthened in that way. There is a natural retentiveness with which some of us are born—the men of colossal intellect—and they remember and are able to use infinitely more things acquired in the past, because they have a brain substance of greater tenacity in holding impressions than others possess. James compares some brains to wax in which the mark left by the seal is permanent; and others he compares to jelly which vibrates at every touch, but retains no dent made in it. From our study of the subconscious we know that the dent did leave an impression on the brain; but it was in the subconscious. So we beg to change the figure and liken, in all mankind, that part of the brain that handles the subconscious to wax, while granting that in some rare cases parts handling the conscious material also hold impressions, as does the wax.

Consequently, according to this theory, we do not strengthen our memories by repetition of facts, lines, or phrases. We cannot grave any deeper the memory paths which nature has provided at birth. But the attention to the thing to be remembered, which repetition has required, has made a larger number of connections of the words with each other, of thought with thought, and of the new with the old. So we have tied the new together with the old by that many more strings, as it were; and any bit of the new tugs at other bits; and the old to which it is tied brings the new with it when it comes to the fore. In other words, careful attention, at the time, to the new stimulus, and its association with the already known, together with repetition, will form a whole system of relations in the mind, and the newly entered material soon become so well-known that it will be difficult to disregard it.

When, in spite of determined effort to remember, the thing is forgotten, especially in the nurse's case, it is usually because the emotional reaction to weariness or to some like obstacle has interfered with proper attention. James advises us if we would improve memory, to improve our thinking processes; to pay more and keener attention, so that we will link things closely together. This in itself will help to arouse interest in the thing to be remembered; and keen interest alone, or careful attention at the time of introduction of the new, and repetition of the thing to be retained, with a will which holds the attention fast, will assure a good, workable memory in any normal mind.

CHAPTER XIII
THE PSYCHOLOGY OF THE NURSE (Continued)

Emotional Equilibrium

Suppose that when you first enter the ward you are wishing with all your heart you had never decided to become a probationer. Perhaps the white screen and its possible meaning has so frightened you that your thoughts refuse to go beyond it. Suppose the very sight of so much sickness has agitated you instead of strengthening your determination to help nurse it. That is, suppose your emotions, your feelings, so fill your mind that perception is necessarily inaccurate and blurred. Then tomorrow your account of the ward will be hazy, and your desire will probably be against returning to a place where so many unpleasant feelings were aroused.

The emotional balance which refuses to allow feelings to obscure judgment by leading reason astray is a necessary safeguard for the work of the nurse. There is little place in the profession for the woman who is "all sentiment," but perhaps there is less for the one without sentiment.

Feeling, we found, is the first expression of mind—feeling which in the early months is entirely selfish. The happiest baby you know is not sweet and winning to please you, but because he feels comfortable and happy and cannot keep from expressing it. His universe is his own little self and you exist only in your relation to him. If you give him pleasure he likes you; if pain, he does not want you. His mother often fails to please him, but satisfies him so much more frequently than anybody else that he loves her best. Then comes nurse or father—if he proves the satisfactory kind of father, or she a nurse he can love. To the baby whatever he happens to want is good. What is not desirable is bad. And such emotional responses are altogether normal in early months, yes, even until the child is old enough to use reason to choose between two desires the one that will in the end prove more satisfying. But they are defects in adult life.

The nurse who would always act as her first feeling dictates would not be in training many days. Unpleasant sights and sounds, the fear of making a mistake which might harm a patient, the undesirability of long hours of hard work in caring for patients who frequently only find fault with her best efforts, would early decide her in favor of another life-work. Comparatively few so-called "grown-ups" are guided only by feeling; and most of those are in institutions that are well safeguarded. But a great many mature men and women allow feeling to unduly influence their thinking. The sentimental nurse, for instance, may find it very difficult to give an ordered hypodermic. The patient dreads the pain and the nurse fears hurting her. Suppose she were to fail to give it on such grounds. This is an almost unthinkable case. But the very nurse who agrees that such an emotional weakling should not be allowed to train, will help her patient, even when recuperating nicely, to grow inexcusably self-centered, by sympathizing with every complaint, warning her at every turn, by allowing her and even encouraging her, perhaps, to discuss her illness and suffering in the minutest detail. This nurse is more damaging than the sentimentalist who fails to give the hypodermic; for that slip is easily discovered, and the transgressor must immediately reform and obey orders, or be dismissed. But the second nurse may take perfect care of the sick body, and the doctor never realize that she is developing the sickness idea in her patient's mind.

In both of these instances reason has followed the leadings of feeling. It is unpleasant to hurt the patient, and she is disagreeable, too, when you insist on carrying out the orders. It is easier to agree with her ideas and sympathize with her troubles, much easier than to find some other avenue for her thinking, or to search for feeling substitutes. It is pleasanter right now to allow her mind to slip unmolested into sick reactions than to lead her, unwilling as she is, into the ways of health. Reason follows feeling's logic, which suggests that it is much better for the patient to talk of her ills than to keep them pent up inside; and judgment is sadly obscured.

The emotionally balanced nurse hears the story once, that she may have the material for helping the need. Feeling, perhaps deep and genuine sympathy with a real trouble, is aroused, and rightly. But this brings a keen desire to help the situation. Reason insists that talking of sufferings, real or fancied, only makes them more insistently felt; that there must be some better way to meet them. It suggests various methods to divert the patient's attention, to change the train of thought until she is able herself to direct it into healthful channels; judgment weighs the propositions and decides upon the one which will lead toward establishing a health attitude.

The nurse is continually meeting the necessity of acting contrary to fear and discouragement and weariness of spirit. How can she secure emotional equilibrium for herself?

Keep in mind the fact that most sick people are very suggestible; that you have a definite responsibility to make your suggestions to your patient wholesome; and that your mood is a constant suggestion to him. Remember that he needs your best. Then, if your own trouble seems too great to bear, determine that, so long as you remain on duty, you will not let it show. Try an experiment. See if you can go through the day carrying your load of sorrow, or disappointment or chagrin, with so serene a face that the sick for whom you are caring will not suspect that you have a burden at all. That is a triumph worth the striving. Then—if you can let it make you a little more comprehending of others' pain, a little more gentle with the sickest ones, a bit more patient with the trying ones, more kindly firm with the unco-operative, realizing that each one of them all has his burden too—you have not choked feeling, but you have fulfilled reason's counsel: that sick people are not the ones to help you in your stress; that a good nurse should rise above personal trouble to the duty at hand. Your judgment has compared your reasons, and decided that you should act before your patients as you would if all were well. And will holds you to emotional equilibrium. Such a thing can be done in a very large measure; and no better opportunity for emotional control will ever be offered than the necessity of being calm and serene before your patients, no matter how you feel.

But, while reason and judgment teach us to control the expression of certain feelings, they urge that this control be exercised in transforming those feelings into helpful ones and giving them an adequate outlet. Such a substitution has been suggested above. Let us not forget that nothing in existence is of personal value until it gives some one an emotion; that feeling is the beauty of life; that living, without the happy, wholesome affective glow, would not be worth the effort; that beauty and strength and sweetness of feeling make for a worthy self. Remember, too, that feeling is the curse of life. It is feeling that would make us give up the whole struggle; and ugliness and weakness and bitterness of feeling make for a despicable self.

Hope lies for us all in the realization that we can choose our feelings, our responses. We can be utterly discouraged, and bitter and depressed at failure; or we can recognize it as a sign-board telling us that the other way than the one we just followed leads to the goal. And we can follow its pointing finger with faith in a new attempt because, now, we know at least how not to go. We can learn despair from all the bitter and the hateful and the mean; or we can learn that they never could be called so if there were not the sweet, the lovable, and the generous with which to compare them. You can learn to search as with a microscope for all the undesirable traits of your patients, or you can calmly accept all that assert themselves as undeniable facts, but use your microscope to find their desirable characteristics which offer possibilities of being brought to the foreground.

You cannot constructively help yourself or your patient by denying the existence of the less worthy traits; but you can resolve to call out the something better. And if you do not find it, as may rarely be the case, you can refuse to let it make you skeptical of finding it in others. Let us remember always that, "It is not things or conditions or people that harm us; it is only the way we respond to them that can hurt." This one great truth, if really believed and made a part of all our thinking, would save scores of people from nervous wreckage. It is a favorite saying of a wise man who has helped a great many people to endure and take new courage when life seemed too hard to meet.

That big, broken-arm case on the ward cursed you yesterday because you would not loosen his splints. And you rushed from the room angry and humiliated, wishing you could quit nursing forever, and asked to be moved because you had been insulted. But that man cannot harm you. He has never known a real lady in his life before. His training from childhood has been to regard women as chattels to do man's bidding; his experience in life is that they usually do what he asks—women of his kind. Moreover, he has never had a serious pain before, and it is not to be endured.

Of course, the man must be dealt with and made to realize the distinction between his new surroundings and the old. Probably the intern or the doctor is the one to do it. Also he must be brought to apologize, or leave the hospital, perhaps. But he did not hurt you. Your own reaction did that. For outside things or people cannot damage what we are in ourselves. The way we respond to them does the harm. When you can control your expression of anger and humiliation, and substitute for your intense feeling a desire that such a patient may learn that pain is often the gateway to healing; that some respect for women may be kindled in him, so that eventually such an outburst in the ward may be impossible for him or for anyone who heard it; then you are choosing between emotions the one of helpfulness, for the one of justified indignation; and feeling has followed reason, rather than leading reason astray. The judgment which decides you to try methods which will shame or inspire some manliness into the patient was one influenced by a well-balanced emotional life.

If we would really acquire emotional poise, there are a few practical, proved methods we might adopt for ourselves.

When we can hold back the expression of the almost overpowering impulse or passion of anger and resentment and hurt; absolutely shut tight our lips until we can think; then wait until we can think without the strain of intense feeling, we will not only keep ourselves out of trouble, but will be able to calmly state our position, right the wrong done us if wrong there was, or recognize that we ourselves were wrong. For we seldom analyze the situation properly under the influence of strong feeling. If we want to accomplish anything with our words, let us wait until we can speak them without having to choke down our sobs or cram back our hot anger, or forcibly restrain ourselves from tearing things or slamming doors. After all that "wild fire" of emotion is gone, judgment will lead us to wisely reasoned action.

Self-correction

Accuracy in work, a primary essential to the nurse, can become automatic if she will demand of herself accuracy of perception, and concentrate on learning and doing until details almost take care of themselves; if she will correct her own work by the standards taught her, and recognize just why and wherein she falls short. Not that she can always do things with the nicety in which they were taught. She cannot give eighteen ward patients in eight hours the same detailed care her private patients would receive if she had only two of them for the same length of time. In such a case she must often sacrifice refinements of detail in service; but there is no excuse for sacrificing accuracy in the necessary treatments of her charges. The nurse merely chooses between the multitude of things which can be done for her ward, the important ones which must be done. Because she is rushed is no excuse for giving a poor hypodermic injection or a careless bed-bath. Accuracy in doing the essential things should be so automatic that it takes not a whit more time than inaccurate doing; and such accuracy is chiefly dependent on constant self-correction when the task is still new, and on never letting up in practice until the details of the doing become practically automatic.

Training the Will

There is no better opportunity for will-training than the hospital affords the nurse. The constant necessity of acting against desire, of doing tasks which in themselves cannot be agreeable, calls for a developed will, while it gives it constant exercise. Moods of discouragement and depression cannot be indulged. The nurse must do her work no matter how tired or blue or "frazzled" she feels, if she is not too sick to be on duty; for all time lost, she knows, is to be made up to the hospital before training is completed.

Can this will to do, despite strong desire to the contrary, this mood control and the ability to disregard physical discomfort, be acquired; and if so, how?

It is a law of the mind and of the body that any task becomes easier by repetition. We found that automatic habit eases much of the strain of action. What seemed repulsive service to the probationer on her first day in the hospital, she forced herself to do because she wanted to be a nurse. She may go on through her three years unreconciled to these particular duties, yet holding herself to them because she likes other features of her work, or because she must earn her living and this seems the best avenue open to her, or because her will to become a nurse is strong enough to make her act continually against desire. And finally, for almost every nurse, the interest in the end to be attained overshadows the unpleasant incidents in its way. The tasks are actually easier by their constant repetition, and her feeling of repugnance becomes only a mild dislike. She has strengthened her will by continuing to act against desire. But there is a better way to the same goal.

The woman who has thought out the reasons for and against taking training; who has considered it carefully as a profession, and has chosen to put up with any obstacles in the way of becoming a graduate nurse, can find a happy adjustment to the disagreeable incidents it involves. Realizing that the paths of learning are seldom thoroughly smooth, she can resolve to use their very roughness for firmer footholds, as a means to self-control, as a fitting for the sterner hardships of self-support, of nursing the dangerously ill, alone, of meeting suffering and death in her patients with quiet courage and faith. In other words, she can meet the thousand and one personal services which in themselves might be disagreeable and prove pure drudgery, not merely with the stern will to do them because they are a necessary part of obtaining a desired end, but also for the sake of adding to the comfort and well-being of each patient in her care. The emotion of interest and kindly desire will ease the strain which will undergoes in demanding that she not shirk the disagreeable. For there is little stress in doing what we wish to do.

It is psychologically possible to find genuine pleasure in the meanest tasks if the doing is backed up by a strong desire to make life count as much for others as possible. The nurse who comes to realize the waste involved in carrying out against desire what reason proposes and volition dictates, will try to secure the co-operation of desire, and save will-force for more worthy accomplishment.

A constant opportunity for will-strengthening comes to many a nurse during the early weeks and months of training in the necessity of going on despite the sheer tiredness, the weary backs and swollen, tender, aching feet. The one who means to "see it through" disregards them as far as possible on duty, gets all the out-of-doors her time permits, takes special exercises to strengthen weak spots, and relaxes her body while she reads or studies or visits in her off-duty time. In the end, not only does her body adjust itself to the new work, but her will has become a better ally for the next demands upon it; her endurance is remarkably increased.

When she can accept hardship, drudgery, weariness of mind and body and perhaps of soul, the nagging of unco-operative patients, and the demands on her sympathies of the suffering; when she can meet these as challenges to develop a strong will—a will not only to endure, but to find happiness and give service through it all—then the nurse has learned the art of making every circumstance a stepping-stone to mastery and achievement.

CHAPTER XIV
THE NURSE OF THE FUTURE

The student of life and of the sciences which deal with the origin and development of the human race, and with the relations of man to man and nation to nation—such sciences as biology and anthropology, sociology and ethics and history—comes to the conclusion that life exists for the development of mind. And mind is not merely intellect, but the only gateway we know to character, to soul. The deepest students of human science see no reason for life except as it "evolves" a perfect mind—man's goal, his ideal. And this visioned perfect mind is one which adjusts itself without friction to the body, making it fulfil the laws of health that it may help and not hinder mind's progress; one which adjusts itself to people and things, co-operating with other minds to develop manners and customs and laws of the most satisfactory community living; one which forces things to be servants of its will; one which makes harmony of life by fulfilling the laws of the soul as well as of the intellect and of the body.

If we believe that life exists for the development of mind into a force of intellect and character and soul, then we need not ask why a nurse should know something of the laws of mind. She does not ask why she should know anatomy or pathology. Her work is dependent upon such knowledge. But if the center of life, the thing which makes the body a living, moving, acting agent instead of a clod, is mind; if the one thing which makes a difference between animal life and mineral and vegetable life is consciousness, i. e., mind; and if everything that affects that body, its organ, affects mind also—then surely no nurse can afford to learn only the rules of repair or of keeping in order the instrument of consciousness, without knowing what effect her efforts have on the mind itself. It is as though an ignorant maid accepted a piano as merely a piece of furniture to be kept clean and shining, and in her zeal to that end scrubbed the keyboard with soap and water which, dripping down into the body of the instrument, swells and damages its felts, rusts and corrodes its keys, and ruins its notes. When she knows that she may thus make impossible the beautiful sounds she has heard it give, and that the more carefully the keyboard is handled the more sure is the beauty resulting, her care is to keep it as free as possible of dust, to see that the top is down and the keyboard covered when she sweeps—and to clean it hereafter in such a way as to never injure its tone.

The nurse has a much greater function than merely to help in saving the body and keeping its machinery in order. If the aim of life is the strengthening and perfecting of the mind—that "urge" of life, then surely the nurse's big aim will be to help establish such health of body as leads toward health of mind. In the average man or woman this vital urge becomes temporarily blocked by the very weakness of the body it urges. The body must give the life-flame some fuel, or it dies out; but with very little fuel it flickers on, waiting, hoping for the more that it may burn strongly again. In the cases the nurse handles very often the "vital spark" has been poorly fed by the disabled body, and so discouragement or depression, or "loss of grip" results, or the flame continues to shine brightly with whatever little sustenance it receives, and so encourages the body to greater effort for it; or sinks into embers, glowing steadily though dully; or it burns wildly, recklessly—it becomes what we call "wild fire," that has no direction and no purpose save to burn up everything it can find.

In other words, the nurse deals with those in whom the "urge" is weakened—the depressed and discouraged; with those whose spirits never flag in their steady shining—those brave souls we could almost worship; and those others who hold grimly on with quiet grit and courage, but with no cheer; and with the unstable ones of neuropathic or psychopathic tendency who become hysteric or maniacal.

What will the nurse do for them all? Will not an understanding of how to recall the ambition to live, the will to get well, and the grit to see the thing through, be an incalculable asset.

The Nurse of the Future

The nurse of the future will not be merely a handmaiden to care for the sick body by deftly carrying out the doctor's orders. She will do this almost automatically as a matter of course, and skilfully; but it will be the merest beginning of her mission. That mission itself will be to eliminate the causes of disease; to teach the ways of health, to supervise the sanitary conditions of city, town, and country. Practical ways and the wise means to this end will be taught in her hospital, which will become a community center with clinics, teaching through its doctors and nurses the way to health, instead of merely treating and advising the cases as they come. But the greatest contribution of the nurse of the future will be a widespread desire for health and will to health, rather than a desire and will to avoid discomfort and pain and danger of death. This will to health will doom in the sane mind the disease-accepting attitude. It will do all that common sense and applied medical science can do to strengthen the body; then it will take what life brings in the way of unavoidable disease and weakness and inability, with an uncringing mind. It will hold the mind's attitude to serenity and poise and accomplishment within the necessary limits of its disordered body. It will be master of its dwelling and make the most of the little the body can give, and force all bearable weakness and pain to be stepping-stones to endurance and will-strength and cheer. It will not accept physical limitations as final things. If life must be lived in a prison-house it will be its own jailer, and fill the rooms with flowers, music, friends, and happiness.

No nurse is competent to help her patient to overcome any curable physical weakness, and keep the mind serene in the face of the incurable, until she herself has learned that the will to health is capable of transforming disease of body, from disaster, into health of mind and soul.

The nurse of the future will know the laws of mind as she knows the course of disease; she will be dedicated to such wise care of existing disease as will lead to prevention of future disease; and she will be a sworn, trained ally of the health-accepting mind.

www.ingramcontent.com/pod-product-compliance
Lightning Source LLC
Chambersburg PA
CBHW011803040426
42450CB00017B/3448